The Iwasawa theory of totally real fields

Ramanujan Mathematical Society Lectures Notes Series

Vol. 1: Number theory

Vol. 2: The Riemann zeta function and related themes

Vol. 3: Formal language aspects of natural computing

Vol. 4: Commutative algebra and combinatorics

Vol. 5: Convexity in discrete structures

Vol. 6: Number theory and discrete geometry

Vol. 7: Discrete mathematics

Vol. 8: Lectures on operator theory

Vol. 9: Essays on geometric group theory

Vol. 10: Techmüller theory and moduli problems

Vol. 11: Perspectives in geometry and topology

Vol. 12: The Iwasawa theory of totally real fields

Vol. 13: Advances in discrete mathematics and applications

Ramanujan Mathematical Society

Lecture Notes Series

Volume 12

The Iwasawa theory of totally real fields

From the workshop on the Iwasawa theory of totally real fields, held at the Indian Institute of Technology, Guwahati, September 2008

Volume editors

J. Coates
C. S. Dalawat
A. Saikia
R. Sujatha

International Press
www.intlpress.com

Ramanujan Mathematical Society
Lecture Notes Series, Volume 12
The Iwasawa theory of totally real fields

Volume editors:
J. Coates
C. S. Dalawat
A. Saikia
R. Sujatha

Published in 2011 by International Press, Somerville, Massachusetts, U.S.A. under license from the Ramanujan Mathematical Society.

This work is also published in the Republic of India, exclusively by the Ramanujan Mathematical Society.

ISBN: 978-1-57146-219-0

Printed in the United States of America.

15 14 13 12 11 1 2 3 4 5 6 7 8 9

Foreword

A Workshop was held at the Indian Institute of Technology, Guwahati, from September 22 until September 30, 2008, with the goal of presenting as much as possible of the background and underlying ideas behind Wiles' celebrated proof of the "main conjecture" of cyclotomic Iwasawa theory over totally real base fields. This workshop was part of a series of schools and workshops organized in various parts of India each year, called "Advanced Training in Mathematics Schools", which are generously supported by the National Board for Higher Mathematics (NBHM). The Organizing Committee consisted of J. Coates, C. S. Dalawat, A. Saikia, and R. Sujatha, and lectures during the course of the Workshop were given by U. K. Anandavardhanan, D. Banerjee, J. Coates, C. S. Dalawat, Narasimha Kumar, E. Ghate, F. A. E. Nuccio, A. Saikia, T. Schmidt, A. C. Sharma, R. Sujatha and O. Venjakob.

We owe to Iwasawa the discovery of the first "main conjecture" in his revolutionary work on cyclotomic fields in the 1960's. He only considered this "main conjecture" for the Tate motive over the field obtained by adjoining all p-power roots of unity to the rational field $\mathbb{Q}$, where p is any prime number. However, it was quickly realized that similar ideas could be applied to elliptic curves, and provided a systematic framework for studying the conjecture of Birch and Swinnerton-Dyer, and other conjectural exact formulae in number theory. Today, we believe, although only fragmentary proofs are known so far, that these "main conjectures" hold in vast generality for all motives over p-adic Lie extensions of number fields. The great interest of the method discovered by Wiles for proving the "main conjecture" for Tate motives over any totally real number field is that it combines ideas from the theory of automorphic forms with purely arithmetic ideas in a way that suggests the beginning of some deeper general connections between these two vast citadels of modern number theory.

These notes present written versions of most of the material covered by the lectures in the Workshop, and it is hoped that they will be useful to others who wish to study these important questions. They are aimed very much at the beginner in the subject, and try to minimize wherever possible the background knowledge required.

In conclusion, we wish to warmly thank IIT Guwahati for providing such excellent facilities and such a delightful location for holding the Workshop.

J. Coates
C. S. Dalawat
A. Saikia
R. Sujatha

Members of the Advisory Board

Contents

Guwahati Workshop on Iwasawa Theory of Totally Real Fields No. 12, 2010, pp. 1–14.

Classical Modular Forms and Galois Representations

T. Schmidt

Mathematisches Institut, Fachbereich Mathematik und Informatik der Universität Münster, Einsteinstrasse 62, 48149 Münster, Deutschland
e-mail: toschmid@math.uni-muenster.de

Introduction

These notes are intended to give a quick survey on classical modular forms for SL_2 and their associated Galois representations. Accordingly, the focus is rather on results and examples whereas proofs are mainly sketched or omitted. Nothing in these notes is new or original and most of the exposed material can be found in any standard textbook on the subject (such as [4], [6], [7]). Where this is not the case we have indicated references.

1. Classical modular forms

1.1 *Definitions and notations*

Throughout these notes we fix two integers $k, N \geq 1$. The symbol p will always denote a prime number. Let $\Gamma(N)$ be the kernel of the reduction map

$$SL_2(\mathbb{Z}) \to SL_2(\mathbb{Z}/N\mathbb{Z}).$$

By a *congruence subgroup* we mean a subgroup of $SL_2(\mathbb{Z})$ containing $\Gamma(N)$ for suitable N. Prominent examples are given by the congruence subgroups *of Hecke type*

$$SL_2(\mathbb{Z}) \supseteq \Gamma_0(N) \supseteq \Gamma_1(N) \supseteq \Gamma(N)$$

where a matrix $\left(\begin{smallmatrix} a & b \\ c & d \end{smallmatrix}\right) \in SL_2(\mathbb{Z})$ lies in $\Gamma_0(N)$ (resp. $\Gamma_1(N)$) iff $c \equiv 0\ (N)$ (resp. $c \equiv 0\ (N),\ a \equiv d \equiv 1\ (N)$). Let $GL_2^+(\mathbb{R})$ be the subgroup of $GL_2(\mathbb{R})$ consisting of matrices with positive determinant. Let $\mathbb{H}$ be the complex upper half plane endowed with its natural $GL_2^+(\mathbb{R})$-action by linear fractional transformations: $\gamma \cdot z := \frac{az+b}{cz+d}$ where a, b, c, d are the entries of γ as above and $z \in \mathbb{H}$. If f is any $\mathbb{C}$-valued function on $\mathbb{H}$ we put

$$(f|_k\gamma)(z) := \det(\gamma)^{k/2}(cz+d)^{-k} f(\gamma \cdot z).$$

This defines a $GL_2^+(\mathbb{R})$-action on the vector space of such functions, the so-called *weight k action*. Let Γ be a fixed congruence subgroup containing $\Gamma(N)$. A *modular form of weight k on* Γ is a complex-valued function on $\mathbb{H}$ satisfying

(a) $f|_k\gamma = f$ for all $\gamma \in \Gamma$,
(b) f is holomorphic on $\mathbb{H}$,
(c) $f|_k\alpha$ is holomorphic at ∞ for all $\alpha \in SL_2(\mathbb{Z})$.

To illustrate the last condition let $SL_2(\mathbb{Z}) = \sqcup_j \Gamma\gamma_j SL_2(\mathbb{Z})_\infty$ be a double coset decomposition where $SL_2(\mathbb{Z})_\infty$ denotes the stabilizer of ∞ in $SL_2(\mathbb{Z})$. Assume condition (b) holds. Then (c) is tantamount to requiring that the finitely many functions $f|_k\gamma_j$ admit power series expansions around ∞ as $\sum_{n\geq 0} a_n q^{n/N}$ with $q = \exp(2\pi i z)$.

Each γ_j as above will take the point ∞ to a point in $\mathbb{Q}\cup\infty$ which is called a *cusp* of Γ. Since Γ has finite index in $SL_2(\mathbb{Z})$ there are *only finitely many cusps*. If f vanishes at all cusps or equivalently if $a_0 = 0$ for the Fourier expansion of f around every $\gamma_j(\infty)$ we call f a *cusp form*. We denote the complex vector spaces of modular forms (resp. cusp forms) of weight k on Γ by $M_k(\Gamma)$ (resp. $S_k(\Gamma)$). Clearly $S_k(\Gamma) \subseteq M_k(\Gamma)$ and if $-1 \in \Gamma$ then (a) implies that $M_k(\Gamma) = 0$ for odd k.

1.2 *Modular curves*

Fix a congruence subgroup Γ. Recall the natural $GL_2(\mathbb{C})$-action on $\mathbb{P}^1(\mathbb{C})$ via

$$\gamma \cdot [x, y] := [ax + by, cx + dy]$$

where $[x, y]$ are homogeneous coordinates. The map

$$\mathbb{H} \longrightarrow \mathbb{P}^1(\mathbb{C}),\ z \mapsto [z, 1]$$

is a $GL_2^+(\mathbb{R})$-equivariant embedding and we put $\mathbb{H}^* := \mathbb{H} \cup \mathbb{P}^1(\mathbb{Q})$. Clearly Γ stabilizes $\mathbb{H}^*$ and we may define the *modular curves*

$$Y(\Gamma) := \Gamma\backslash\mathbb{H}, \quad X(\Gamma) := \Gamma\backslash\mathbb{H}^*.$$

Mapping $\alpha \mapsto \alpha(\infty)$ induces a bijection

$$\Gamma\backslash SL_2(\mathbb{Z})/SL_2(\mathbb{Z})_\infty \xrightarrow{\cong} X(\Gamma) - Y(\Gamma).$$

Thus, adding the cusps of Γ to $Y(\Gamma)$ yields $X(\Gamma)$ which in fact is compact. The space $X(\Gamma)$ can be endowed with the structure of a complex Riemann surface where suitable charts are given (around most points) via the natural projection $\mathbb{H} \to X(\Gamma)$. Applying the Riemann-Roch theorem to $X(\Gamma)$ yields precise formulas (except for $k = 1$) for the dimensions of $M_k(\Gamma)$ and $S_k(\Gamma)$. In particular,

Theorem 1.1. $\dim_{\mathbb{C}} M_k(\Gamma) < \infty$.

Following common usage we write $X(N)$, $X_0(N)$, $X_1(N)$ instead of $X(\Gamma)$ in case of $\Gamma = \Gamma(N), \Gamma_0(N), \Gamma_1(N)$ respectively.

1.3 *Example: Level 1*

A fundamental domain for the action of $SL_2(\mathbb{Z})$ on $\mathbb{H}$ is given by

$$\{z \in \mathbb{H} : |z| \geq 1 \text{ and } -1/2 \leq Re(z) \leq 1/2\}.$$

Hence, $SL_2(\mathbb{Z})$ has only one cusp namely ∞. In fact, it can be shown that there is a complex analytic isomorphism $X(1) \xrightarrow{\cong} \mathbb{P}^1(\mathbb{C})$ from which one may deduce

Theorem 1.2. *Denoting by $\lfloor \cdot \rfloor$ the function "greatest integer" one has*

$$\dim_{\mathbb{C}} M_k(SL_2(\mathbb{Z})) = \begin{cases} 0 & k = 2 \\ \lfloor k/12 \rfloor & k \equiv 2\ (12), k > 2 \\ \lfloor k/12 \rfloor + 1 & \textit{else.} \end{cases}$$

Since $f \in M_k(SL_2(\mathbb{Z}))$ is a cusp form iff it vanishes at infinity there is an exact sequence for even $k \geq 4$

$$0 \longrightarrow S_k(SL_2(\mathbb{Z})) \longrightarrow M_k(SL_2(\mathbb{Z})) \longrightarrow \mathbb{C} \longrightarrow 0$$

induced by the map $\sum_{n \geq 0} a_n q^n \to a_0$. It follows that in this case there is a unique normalized modular form

$$E_k := 1 + a_1 q + a_2 q^2 + \cdots$$

inside $M_k(SL_2(\mathbb{Z}))$, the classical *Eisenstein series of weight* k on $SL_2(\mathbb{Z})$. Its very existence is due to complex lattice theory which we will sketch in the next section.

1.4 *Complex lattices and elliptic curves*

There is a close relation between modular forms and elliptic curves. To illustrate this, we will briefly talk about modular forms of level 1 and elliptic curves with complex coefficients. By a *lattice* in $\mathbb{C}$ we mean a discrete subgroup of the form $\mathbb{Z}w_1 + \mathbb{Z}w_2$ where the $w_i \in \mathbb{C}$ are linearly independent over $\mathbb{R}$ and $w_1/w_2 \in \mathbb{H}$. Let $\mathcal{L}$ be the set of all lattices in $\mathbb{C}$. There is then a $\mathbb{C}$-linear isomorphism between

$$M_1'(SL_2(\mathbb{Z})) := \{f : \mathbb{H} \to \mathbb{C},\ f|_k\gamma = f \text{ for all } \gamma \in SL_2(\mathbb{Z})\},$$
$$L_1'(SL_2(\mathbb{Z})) := \{F : \mathcal{L} \to \mathbb{C},\ F(\lambda L) = \lambda^{-k} F(L) \text{ for all } \lambda \in \mathbb{C}^\times\}$$

given by

$$f \mapsto F_f(\mathbb{Z}w_1 + \mathbb{Z}w_2) := w_2^{-k} f(w_1/w_2).$$

Denoting by ζ the Riemann ζ-function the series

$$E_k(z) := \frac{1}{2}\zeta(1-k)^{-1} \sum_{w \in \mathbb{Z}z + \mathbb{Z}, w \neq 0} w^{-k}$$

for $k \geq 4$ and $z \in \mathbb{H}$ is easily seen to be absolutely and locally uniformly convergent. It therefore defines a holomorphic function of z on $\mathbb{H}$. Clearly, $E_k = 0$ if k is odd and from the above isomorphism we see that $E_k \in M_1'(SL_2(\mathbb{Z}))$. Computing the Fourier expansion of E_k around ∞, one obtains

$$E_k(q) = 1 - \frac{2k}{B_k} \sum_{n \geq 1} \sigma_{k-1}(n) q^n$$

where B_k is the kth Bernoulli number and $\sigma_{k-1}(n) := \sum_{0<d|n} d^{k-1}$. Hence $E_k, k \geq 4$ is the Eisenstein series referred to above.

Given a lattice $L \in \mathcal{L}$ the quotient $\mathbb{C}/L$ is a complex torus which gives rise to an elliptic curve over $\mathbb{C}$ as follows. If $\wp$ denotes the Weierstrass $\wp$-function associated to L, there is an isomorphism (of complex Lie groups)

$$\mathbb{C}/L \longrightarrow E_L(\mathbb{C}),\ z \mapsto [\wp(z), \wp'(z), 1]$$

onto the $\mathbb{C}$-points of an elliptic curve $E_L/\mathbb{C}$ given as a cubic in $\mathbb{P}^2(\mathbb{C})$ by a Weierstrass equation of the form

$$E_L : y^2 = 4x^3 - g_2 x - g_3$$

with constants $g_i = g_i(L)$. Under this isomorphism the holomorphic and nowhere vanishing invariant differential dz is mapped to dx/y. This sets up a bijection

$$\mathcal{L} \cong \{(E, \omega) : E/\mathbb{C} \text{ and } \omega \in \Omega_E \text{ holomorphic nonvanishing}\}$$

given by $L \mapsto (E_L, dx/y)$. This is the well known uniformization for complex elliptic curves. In this way a modular form $f \in M_k(SL_2(\mathbb{Z}))$ may be regarded as a "rule" $\mathbb{F}$ assigning each pair (E, ω) as above an element $F_f(L(E, \omega)) \in \mathbb{C}$ in a "holomorphic" way and with the property that $\mathbb{F}((E, \lambda\omega)) = \lambda^{-k}\mathbb{F}((E, \omega))$ for all $\lambda \in \mathbb{C}^\times$. Along these lines the theory of analytic modular forms generalizes vastly to so-called "geometric modular forms", a theory for which we refer to [5].

1.5 *The Petersson inner product*

Let Γ be a congruence subgroup. The differential

$$dv := \frac{dx \wedge dy}{y^2}$$

on $\mathbb{H}$ is easily seen to be $SL_2(\mathbb{Z})$-invariant and therefore extends to $Y(\Gamma)$. Let $f, g \in M_k(\Gamma)$ and let one of them be a cusp form. The function on $\mathbb{H}$

$$\delta(f, g)(z) := f(z)\overline{g(z)}\mathrm{Im}(z)^k$$

is Γ-invariant in the sense

$$\delta(f(\gamma \cdot z), g(\gamma \cdot z)) = \delta(f, g)(z)$$

for $\gamma \in \Gamma$, $z \in \mathbb{H}$. Here, we denote by $\mathrm{Im}(z)$ the imaginary part of z and use $\mathrm{Im}(\gamma \cdot z) = \mathrm{Im}(z)/|cz+d|^2$. Since $(fg)(\infty) = 0$, the absolute value $|\delta(f,g)|$ is bounded on $\mathbb{H}$. From this it is not hard to see that the integral

$$(f,g) := v(Y(\Gamma))^{-1} \int_{Y(\Gamma)} \delta(f,g)\, dv$$

is well-defined and gives an hermitian inner product on $S_k(\Gamma)$. One verifies immediately that if $\Gamma' \subseteq \Gamma$ is another congruence subgroup then $(f,g)_\Gamma = (f,g)_{\Gamma'}$. It is therefore not ambiguous to omit the level from the notation. Furthermore if $\gamma \in SL_2(\mathbb{Z})$ then one may deduce that

$$(f|_k\gamma, g|_k\gamma) = v(Y(\Gamma))^{-1} \int_{Y(\Gamma)} \delta(f,g) \circ \gamma \; dv = (f,g).$$

Summing up:

Theorem 1.3. *$(\cdot,\cdot)$ is an $SL_2(\mathbb{Z})$-invariant hermitian inner product on $S_k(\Gamma)$.*

The orthogonal complement $N_k(\Gamma) := S_k(\Gamma)^\perp$ inside $M_k(\Gamma)$ is called the *space of Eisenstein series of level* Γ. The name is justified by the following construction. For simplicity we let $\Gamma := \Gamma(N)$ and $k \geq 3$. For integers $0 \leq \mu, \nu < N$ put

$$E_k(z,\mu,\nu,N) := \sum_{m,n} (mz+n)^{-k}$$

where the summation runs over $m \equiv \mu\ (N)$ and $n \equiv \nu\ (N)$ omitting the point $(0,0)$ in case $\mu = \nu = 0$. Clearly, $E_k(z,0,0,1) = E_k$ whence $E_k(z,\mu,\nu,N)$ enjoys the same convergence properties as E_k and is therefore holomorphic on $\mathbb{H}$. Each $E_k(z,\mu,\nu,N)$ is even a modular form: condition (a) is a short computation with congruences and condition (b) follows from the two facts: (i) the $\mathbb{C}$-span

$$\sum_{0 \leq \nu,\mu < N} \mathbb{C}\, E_k(z,\mu,\nu,N)$$

is $SL_2(\mathbb{Z})$-stable and (ii) each element in it is holomorphic at ∞. The latter follows, as in the case of E_k, by explicitly computing q-expansions. The elements $E_k(z,\mu,\nu,N)$ are called *Eisenstein series of higher level*.

Theorem 1.4. *One has*

$$N_k(\Gamma(N)) = \sum_{0 \leq \nu,\mu < N} \mathbb{C}\, E_k(z,\mu,\nu,N).$$

1.6 *Diamond operators and Nebentypus*

We start with some preliminaries on the level. Let Γ be a congruence subgroup. Since $\Gamma \supseteq \Gamma(N)$, some N, we have $M_k(\Gamma) \subseteq M_k(\Gamma(N))$. Furthermore, letting

$$\delta_N := \begin{pmatrix} N & 0 \\ 0 & 1 \end{pmatrix} \in GL_2^+(\mathbb{R})$$

one has $\delta_N^{-1}\Gamma(N)\delta_N \supseteq \Gamma_1(N^2)$. Hence, given $f \in M_k(\Gamma(N))$ we obtain

$$f(Nz) = f(\delta_n \cdot z) = N^{-k/2}(f|_k\delta_N) \in M_k(\delta_N^{-1}\Gamma(N)\delta_N) \subseteq M_k(\Gamma_1(N^2)).$$

If furthermore $f = \sum_n a_n q^{n/N}$ equals the Fourier expansion of f then $f(Nz) = \sum_n a_n q^n$. Hence, by "conjugation" we may view f as having level $\Gamma_1(N^2)$ without changing its Fourier coefficients. This is why we will restrict to level $\Gamma_1(N)$ in the following. Note also that $f \in M_k(\Gamma(N))$ actually lies in $M_k(\Gamma_1(N))$ iff $f(z) = f(z+1)$. We will reduce our analysis of the space $M_k(\Gamma_1(N))$ even one step further. There is a group isomorphism

$$\Gamma_0(N)/\Gamma_1(N) \xrightarrow{\cong} (\mathbb{Z}/N)^\times, \begin{pmatrix} a & b \\ c & d \end{pmatrix} \mapsto d.$$

We may therefore unambiguously define for given $d \in (\mathbb{Z}/N)^\times$, $f \in M_k(\Gamma_1(N))$

$$f|\langle d\rangle := f|_k\gamma$$

where $\gamma \in \Gamma_0(N)$ is any matrix having d in its lower right corner. We call $\langle d\rangle$ the *diamond operator* associated with $d \in (\mathbb{Z}/N)^\times$. It preserves the subspace of cusp forms. Given a *Dirichlet character mod N* i.e. a homomorphism

$$\varepsilon : (\mathbb{Z}/N)^\times \longrightarrow \mathbb{C}^\times$$

define

$$M_k(N,\varepsilon) := \{f \in M_k(\Gamma_1(N)),\ f|\langle d\rangle = \varepsilon(d)f \ \text{ for all } \ d \in (\mathbb{Z}/N)^\times\}$$

and $S_k(N,\varepsilon) := M_k(N,\varepsilon) \cap S_k(\Gamma_1(N))$. Elements in these spaces are said to have *Nebentypus* ε. Note that if ε has not the same parity as k (i.e. $\varepsilon(-1) \neq (-1)^k$) then $-1 \in \Gamma_0(N)$ implies $M_k(N,\varepsilon) = 0$. By semisimplicity we now have

$$M_k(\Gamma_1(N)) = \oplus_\varepsilon M_k(N,\varepsilon)$$

and similarly for $S_k(N,\varepsilon)$ where ε runs through all Dirichlet characters mod N. This decomposition is orthogonal with respect to the Petersson inner product.

Proposition 1.5. *If $\varepsilon_1 \neq \varepsilon_2$ then $(f_1, f_2) = 0$ for $f_i \in M_k(N,\varepsilon_i)$.*

Indeed, pick $\gamma \in \Gamma_0(N)$ such that $\varepsilon_1(\gamma) \neq \varepsilon_2(\gamma)$. Then

$$\varepsilon_1(\gamma)(f_1, f_2) = (f_1|_k\gamma, f_2) = (f_1, f_2|_k\gamma^{-1}) = \varepsilon_2(\gamma)(f_1, f_2)$$

by the $SL_2(\mathbb{Z})$-invariance of $(\cdot,\cdot)$.

We will henceforth restrict to a fixed Nebentypus ε.

1.7 *Hecke operators*

The theory of modular forms is enriched by the presence of Hecke operators. We first give an ad-hoc definition and relate it afterwards to the usual double coset formalism.

Let $n \geq 1$ and define for $f \in M_k(N, \varepsilon)$, $z \in \mathbb{H}$

$$(f|T_n)(z) := n^{k-1} \sum_{ad=n} \sum_{b=0}^{d-1} \varepsilon(a) d^{-k} f\left(\frac{az+b}{d}\right)$$

with $\varepsilon(a) := 0$ if $(a, N) \neq 1$.

Theorem 1.6. *Each T_n defines an operator on $M_k(N, \varepsilon)$ stabilizing $S_k(N, \varepsilon)$.*

T_n is called the n-th *Hecke operator* on $M_k(N, \varepsilon)$. Define the *Hecke algebra* $\mathcal{H}(N, k, \varepsilon)$ for $M_k(N, \varepsilon)$ to be the $\mathbb{C}$-subalgebra of $\mathrm{End}_{\mathbb{C}}(M_k(N, \varepsilon))$ generated by the elements T_p, $\langle p \rangle$ for $p \nmid N$ and by $U_p := T_p$ for $p|N$. It has finite dimension over $\mathbb{C}$. The U_p-operators are sometimes called *Atkin-Lehner operators* due to their significance in Atkin-Lehner theory (cf. [1]).

Theorem 1.7. *$\mathcal{H}(N, k, \varepsilon)$ is commutative and contains all $T_n, n \geq 1$.*

The commutativity follows from a general double coset formalism which we will sketch below together with the presence of certain anti-involutions on representatives of these double cosets. The fact that $\mathcal{H}(N, k, \varepsilon)$ contains in fact all $T_n, n \geq 1$ follows from certain recursion formulas. For example, one has

$$T_p T_{p^e} = \begin{cases} T_{p^{e+1}} + p^{k-1} \langle p \rangle T_{p^{e-1}} & p \nmid N \\ T_{p^{e+1}} & p|N \end{cases}$$

and similarly $T_n T_m = T_{nm}$ if $(n, m) = 1$.

Theorem 1.8. *Given $f \in M_k(N, \varepsilon)$ with q-expansion $\sum_n a_n q^n$ one has*

$$(f|T_p)(q) = \sum_n a_{pn} q^n + \varepsilon(p) p^{k-1} \sum_n a_n q^{pn},$$

$$(f|U_p)(q) = \sum_n a_{pn} q^n$$

Note that the first formula implies the second since $\varepsilon(p) = 0$ if $(p, N) \neq 1$.

Remark. Invoking the modular curve $X_1(N)$ the Hecke operators $T_n, n \geq 1$ can be given an interpretation as algebraic correspondences on the smooth algebraic curve associated via GAGA to the compact Riemann surface $X_1(N)$. As such they induce endomorphisms of the Jacobian variety of $X_1(N)$. We will not pursue this point of view any further (e.g. [7], §2.8).

Recall that $(\cdot, \cdot)$ makes $S_k(N, \varepsilon)$ a unitary space whence each $T_n, n \geq 1$ has an adjoint. For primes outside the level the latter are easy to determine.

Theorem 1.9. *Let $(p, N) = 1$. Then $T_p^* = \bar{\varepsilon} T_p$ and $\langle p \rangle^* = \langle p \rangle^{-1}$. In particular, $S_k(N, \varepsilon)$ has a basis consisting of simultaneous eigenvectors for all these operators.*

Proof. The first identity follows from the $SL_2(\mathbb{Z})$-invariance of $(\cdot, \cdot)$ and the second is immediate. Hence all these operators are normal whence the last statement follows since $\mathcal{H}(N, k, \varepsilon)$ is commutative.

A cusp form $f = \sum_{n\geq 1} a_n q^n \in S_k(N, \varepsilon)$ is called a *normalized eigenform* if $T_n(f) = \lambda_n f$ with $\lambda_n \in \mathbb{C}$ for all $n \geq 1$ and $a_1 = 1$. In this case, $\lambda_n = a_n$ and the subfield $K_f = \mathbb{Q}(a_n, n \geq 1) \subseteq \mathbb{C}$ is called the *Hecke field* of f. It is a finite extension of $\mathbb{Q}$ containing the values $\varepsilon(p)$ for $p \nmid N$.

We conclude with the important fact that the formal Dirichlet series associated to the $T_n, n \geq 1$ has an Euler product. Let us make this precise. Let A be any commutative ring and form the formal power series ring $\tilde{A} := A[[u_p : p \text{ prime}]]$ in the infinitely many variables u_p. Elements of $\tilde{A}$ are called *formal Dirichlet series* over A. Writing $n^{-s} := \prod_p u_p^{e_p}$ if $n = \prod_p p^{e_p} \geq 2$ and $1^{-s} := 1$ we see that any element in $\tilde{A}$ has a unique expression as $\sum_{n\geq 1} a_n n^{-s}$ with a sequence $\{a_n\}$ in A. Let $\phi_p(s) \in A[[u_p]]$ be given for all primes p. Suppose $\phi_p(s) \in 1+u_p\, A[[u_p]]$ for almost all p. The product $\prod_p \phi_p(s)$ is then meaningful in $\tilde{A}$. If an arbitrary element ϕ in $\tilde{A}$ can be expressed as such a product we will say that ϕ has a *formal Euler product*.

Applying this to the commutative ring $\mathcal{H}(N, k, \varepsilon)$ one may prove the

Theorem 1.10. *The formal Dirichlet series $\sum_n T_n n^{-s}$ has the formal Euler product*

$$\prod_{p\mid N}(1 - T_p p^{-s})^{-1} \times \prod_{p\nmid N}(1 - T_p p^{-1} + p^{k-1}\langle p\rangle p^{-2s})^{-1}.$$

1.8 *Interlude: Double coset formalism*

To introduce Hecke algebras in a general automorphic setting one usually invokes a certain double coset formalism. We sketch it here in our case of interest. Let

$$\Delta_0(N) := \left\{ \begin{pmatrix} a & b \\ c & d \end{pmatrix} \in M_2(\mathbb{Z}), c \equiv 0\ (N), (a, N) = 1, ad - bc > 0 \right\}.$$

It is a semi-group containing $\Gamma_0(N)$. Abbreviate

$$[\alpha] := \Gamma_0(N)\alpha\Gamma_0(N)$$

for the double coset of $\alpha \in \Delta_0(N)$ inside $\Gamma_0(N)\backslash GL_2^+(\mathbb{R})/\Gamma_0(N)$. Let $R(N)$ be the free $\mathbb{Z}$-module on symbols $[\alpha], \alpha \in \Delta_0(N)$ endowed with the following multiplication: let $[\alpha], [\beta]$ be given. By our choice of $\Delta_0(N)$ there are decompositions into left cosets

$$[\alpha] = \bigsqcup_i \Gamma_0(N)\alpha_i, \quad [\beta] = \bigsqcup_j \Gamma_0(N)\beta_j$$

with *finitely* many $\alpha_i, \beta_j \in \Delta_0(N)$ (note that $[\Gamma_0(N) : \Gamma_0(N) \cap \alpha^{-1}\Gamma_0(N)\alpha] < \infty$, cf. [7], Lem. 2.7.1). Then we put

$$[\alpha] \cdot [\beta] := \sum_\gamma c_\gamma [\gamma]$$

where the summation runs through all double cosets $[\gamma] \subseteq \Delta_0(N)$ and $c_\gamma := \#\{(i, j) : \Gamma_0(N)\alpha_i\beta_j = \Gamma_0(N)\gamma\}$. It is clear that $c_\gamma = 0$ for almost all γ whence extending this definition linearly gives an associative unital $\mathbb{Z}$-algebra structure on $R(N)$. Crucial is the following

Lemma 1.11. *Any $\alpha \in \Delta_0(N)$ determines uniquely positive integers $l, m \geq 1$ such that $l|m$, $(l, N) = 1$ and*

$$[\alpha] = \Gamma_0(N) \begin{pmatrix} l & 0 \\ 0 & m \end{pmatrix} \Gamma_0(N)$$

as double cosets inside $\Gamma_0(N)\backslash GL_2^+(\mathbb{R})/\Gamma_0(N)$.

Define

$$T(l, m) := \left[\begin{pmatrix} l & 0 \\ 0 & m \end{pmatrix} \right]$$

for all such (l, m) and $T(n) := \sum_{lm=n} T(l, m)$ for all $n \geq 1$.

Theorem 1.12. *The algebra $R(N)$ equals the polynomial ring over $\mathbb{Z}$ in the variables $T(p)$, $T(p, p)$, $p \nmid N$ and $T(q)$, $q|N$ where q, p are prime numbers.*

To connect this approach to the former we define an $R(N)$-action on $M_k(N, \varepsilon)$ as follows. The character ε extends from $\Gamma_0(N)$ to $\Delta_0(N)$ via

$$\varepsilon\left(\begin{pmatrix} a & b \\ c & d \end{pmatrix}\right) := \bar{\varepsilon}(a)$$

(note that if $\gamma \in \Gamma_0(N)$ then $ad \equiv 1$ (N) implies $\bar{\varepsilon}(a) = \varepsilon(d)$). Let $\alpha \in \Delta_0(N)$ and write $[\alpha] = \sqcup_i \Gamma_0(N)\alpha_i$ with finitely many $\alpha_i \in \Delta_0(N)$. Then if $f \in M_k(N, \varepsilon)$ define

$$f|[\alpha] := \det(\alpha)^{k/2-1} \sum_i \bar{\varepsilon}(\alpha_i)(f|_k\alpha_i).$$

We obtain well defined operators on $M_k(N, \varepsilon)$ that stabilize $S_k(N, \varepsilon)$.

Theorem 1.13. *The ring homomorphism $R(N) \longrightarrow \mathrm{End}_{\mathbb{C}}(M_k(N, \varepsilon))$ satisfies $T(p) \mapsto T_p$, $T(p, p) \mapsto p^{k-2}\langle p\rangle$ and $T(q) \mapsto U_q$.*

1.9 *L-series attached to modular forms*

Let $f = \sum_{n\geq 1} a_n q^n \in S_k(N, \varepsilon)$, $q = \exp(2\pi i z)$. By standard Fourier theory

$$a_n = \int_0^1 \exp(-2\pi i n(x + iy)) f(x + iy) dx$$

for any $y \in \mathbb{R}$. Since $f(\infty) = 0$, it is easy to see that $y^{k/2}|f|$ is bounded on $\mathbb{H}$ whence we obtain

$$|a_n| \ll y^{-k/2} \exp(2\pi n y) \ll n^{k/2}$$

by putting $y := 1/n$. It follows that the Dirichlet series

$$L(s, f) := \sum_{n\geq 1} a_n n^{-s}, \ s \in \mathbb{C}$$

converges absolutely and locally uniformly in the strip $Re(s) > k/2 + 1$ and defines a holomorphic function. Recall that the complex Γ-function $\Gamma(s)$ is a meromorphic function on $\mathbb{C}$ with no zeroes whose poles are at the non-positive integers and simple. For $Re(s) > 0$ it admits the explicit description

$$\Gamma(s) = \int_0^{\infty} t^{s-1} \exp(-t)\, dt.$$

Theorem 1.14. *The function* $\Lambda_N(s, f) := (2\pi/\sqrt{N})^{-s}\Gamma(s)L(s, f)$ *has analytic continuation to* $\mathbb{C}$ *and satisfies the functional equation*

$$\Lambda_N(s, f) = i^k \Lambda_N(k - s, f|_k\, \omega_N)$$

where $\omega_N := \left(\begin{smallmatrix} 0 & -1 \\ N & 0 \end{smallmatrix}\right)$.

It follows that $L(s, f)$ is an *entire* function on $\mathbb{C}$ with simple zeroes at the non-positive integers. When has it an Euler product?

Theorem 1.15. *Let* $f \in S_k(N, \varepsilon)$ *be nonzero. The following conditions are equivalent:*

1. f *is an eigenform for all* $T_n, n \geq 1$.
2. $a_1 \neq 0$ *and* $L(s, f) = a_1 \prod_p (1 - t(p)p^{-s} + \varepsilon(p)p^{k-1-2s})^{-1}$.

In this case $f|T_n = t(n)f$ *with* $t(n) = a_n/a_1$.

Proof. This follows from the formal Euler product over $\mathcal{H}(N, k, \varepsilon)$ appearing in Thm. 1.10.

Example. let $N = 1, k = 12$ and recall *Ramanujan's* Δ*-function*

$$\Delta := (E_4^3 - E_3^4)/12^3 \in S_{12}(SL_2(\mathbb{Z})).$$

Since $\dim_{\mathbb{C}} S_{12}(SL_2(\mathbb{Z})) = 1$ by Thm. 1.2 the function Δ has to be an eigenform for all T_n. We denote (for historical reasons) its q-expansion by $\sum_{n\geq 1} \tau_n q^n$. Since $\tau_1 = 1$ we obtain from the theorem

$$L(s, \Delta) = \prod_p (1 - \tau_p p^{-s} + p^{11-2s})^{-1}.$$

In this situation it is interesting to know that $|\tau_p| \leq 2p^{11/2}$ since this used to be the famous *Ramanujan conjecture* (proved by Deligne). Furthermore, all coefficients $\tau_n, n \geq 1$ are in $\mathbb{Z}$ and Ramanujan proved many astonishing arithmetic properties about them e.g. $\tau_n \equiv \sigma_{11}(n)$ (691).

1.10 *Newforms*

We have seen that $S_k(N, \varepsilon)$ has a basis consisting of eigenforms f for all T_p where p is prime to the level. However, the corresponding L-series $L(s, f)$ admit usually no Euler product due to the missing factors in the level (cf. Thm. 1.15). The idea to remedy this asymmetry is to restrict to a subspace of $S_k(N, \varepsilon)$ (obtained essentially by exluding all cusp forms coming from a lower level) over which the *whole* Hecke algebra is diagonalizable. To make this precise let $l \geq 1$ be an integer and put $\delta_l := \left(\begin{smallmatrix} l & 0 \\ 0 & 1 \end{smallmatrix}\right)$ as before. If $f \in M_k(N, \varepsilon)$ then

$$f(lz) = l^{-k/2}(f|_k\delta_l)(z) \in M_k(Nl, \varepsilon|_{\Gamma_0(Nl)})$$

(see above) and similarly for $S_k(N, \varepsilon)$. Let m_ε be the conductor of ε. Let (M, l) be positive integers such that $m_\varepsilon | M | N$ with $M \neq N$ and $l | (N/M)$. Given $f \in S_k(M, \varepsilon)$ we have $f(lz) \in S_k(Ml, \varepsilon) \subseteq S_k(N, \varepsilon)$ where the latter holds since $Ml | N$ and therefore $\Gamma_0(N) \subseteq \Gamma_0(Ml)$. It makes therefore sense to define the *space of old forms*

$$S_k^1(N, \varepsilon)$$

to be the $\mathbb{C}$-span generated inside $S_k(N, \varepsilon)$ by the set

$$\bigcup_M \bigcup_l \{f(lz), f(z) \in S_k(M, \varepsilon)\}$$

where (M, l) runs through the positive integers satisfying the conditions above. The *space of newforms*

$$S_k^0(N, \varepsilon) := S_k^1(N, \varepsilon)^\perp$$

is the orthogonal complement with respect to the Petersson inner product. Of course, it might very well be zero. However, if ε is primitive (i.e. $m_\varepsilon = N$) then clearly $S_k^0(N, \varepsilon) = S_k(N, \varepsilon)$.

Theorem 1.16. *Both spaces $S_k^1(N, \varepsilon)$ and $S_k^0(N, \varepsilon)$ are stable under the induced actions of T_n, $(n, N) = 1$.*

Proof. $S_k^1(N, \varepsilon)$ is stable since the T_n, $(n, N) = 1$ essentially commute with the δ_l where l is as above. Using the formulas for the adjoints (Thm. 1.9) one deduces stability of $S_k^1(N, \varepsilon)^\perp = S_k^0(N, \varepsilon)$.

Theorem 1.17 ("multiplicity one"). *Assume $f \in S_k(N, \varepsilon)$ and $g \in S_k^0(N, \varepsilon)$. Suppose that for each $(n, N) = 1$ the functions f, g are eigenforms of T_n with the same eigenvalue. Then $g \in \mathbb{C} f$.*

The theorem implies that Hecke characters appearing in the eigenspace decomposition of $S_k^0(N, \varepsilon)$ with respect to the T_n, $(n, N) = 1$ have multiplicity one.

An element $f = \sum_n a_n q^n \in S_k^0(N, \varepsilon)$ is called *primitive of conductor* N if it is an eigenform for all T_n, $(n, N) = 1$ and if it is normalized (i.e. $a_1 = 1$).

Theorem 1.18. *Primitive forms are eigenforms for the whole Hecke algebra $\mathcal{H}(N, k, \varepsilon)$. $S_k^0(N, \varepsilon)$ has a basis consisting of primitive forms.*

Proof. Let $f \in S_k^0(N, \varepsilon)$ be primitive and $T \in \mathcal{H}(N, k, \varepsilon)$. Since $\mathcal{H}(N, k, \varepsilon)$ is commutative we may apply the multiplicity one theorem to f and $f | T$ which yields $f | T \in \mathbb{C} f$. The second statement is clear since $S_k(N, \varepsilon)$ has such a basis according to Thm. 1.9.

2. Galois representations

2.1 *l-adic representations*

We briefly recall some background from l-adic representation theory. For more details we refer to the lucid exposition in [8]. Let K be a field, K_s its separable closure and $G_K := Gal(K_s/K)$. Let l be a prime and V a finite dimensional $\mathbb{Q}_l$-vector space. An *l-adic representation* of G_K is a continuous homomorphism

$$G_K \longrightarrow Aut_{\mathbb{Q}_l}(V).$$

Examples.

(1) *Roots of unity*. Let $l \neq char\ K$ and μ_n the group of l^n-th roots of unity in K_s, $T_l(\mu) := \varprojlim_n \mu_n$ the l-adic Tate module and $V_l(\mu) = T_l(\mu) \otimes_{\mathbb{Z}_l} \mathbb{Q}_l$. The natural continuous action of G_K on μ_n induces an l-adic representation

$$\chi_l : G_K \longrightarrow Aut(V_l(\mu)).$$

This is of course nothing else than the l-adic cyclotomic character.

(2) *Elliptic curves*. Let $l \neq char\ K$, E be an elliptic curve over K and E_n the group of l^n-torsion points of E. Let $T_l(E) := \varprojlim_n E_n$ be the l-adic Tate module and $V_l(E) = T_l(E) \otimes_{\mathbb{Z}_l} \mathbb{Q}_l$. The natural continuous action of G_K on E_n induces an l-adic representation

$$\pi_l : G_K \longrightarrow Aut(V_l(E)).$$

The image of π_l is a closed subgroup isomorphic to $GL_2(\mathbb{Z}_l)$, i.e. an l-adic Lie group. Duality of abelian varieties induces a canonical isomorphism $\wedge^2 V_l(E) \cong V_l(\mu)$ whence $\det(\pi_l) = \chi_l$.

Remark. Replacing l and $\mathbb{Q}_l$ by a finite place λ of a number field $L/\mathbb{Q}$ and the completion L_λ one defines *λ-adic representations* in an analogous manner.

From now on K will be a number field. Let $\rho : G_K \to Aut(V)$ be a λ-adic representation and v a finite place of K. ρ is said to be *unramified* at v if $\rho(I_w) = 1$ for the inertia subgroup $I_w \subseteq G_K$ of every extension w of v to K_s. Let D_w be the decomposition group and $F_w \in D_w/I_w$ a Frobenius element at $w|v$. If ρ is unramified at v its restriction $\rho|_{D_w}$ factors through D_w/I_w whence $\rho(F_w)$ is well defined. Since decomposition group and inertia are unique up to conjugation the conjugacy class of this element in $Aut(V)$ depends only on v and we denote it by $\rho(Frob_v)$. For example, in the above examples χ_l is unramified outside l and π_l is unramified outside l and places of bad reduction of E.

We record for future reference the following simple application of the *Čebotarev density theorem* (cf. [8], I.§2 Cor. 2).

Proposition 2.1. *Let L/K be a Galois extension (finite or not) of the number field K, unramified outside a finite set of places S. The elements F_w, $w|v$, $v \notin S$ form a dense subset of $Gal(L/K)$.*

2.2 *Modular forms and λ-adic representations*

Let as above $k, N \geq 1$ and ε a Dirichlet character mod N with $\varepsilon(-1) = (-1)^k$. Primitive forms give rise to adic Galois representations in the following sense.

Theorem 2.2 (Deligne-Eichler-Serre-Shimura). *Let*

$$f = \sum_n a_n q^n \in S_k^0(N, \varepsilon)$$

be a primitive form of conductor N. Let $K/\mathbb{Q}$ be the Hecke field of f. Let λ be a finite place of K, K_λ the completion and l the characteristic of its residue field. There exists a λ-adic representation

$$\rho_{f,\lambda} : G_{\mathbb{Q}} \longrightarrow GL_2(K_\lambda)$$

with the following properties:

1. $\rho_{f,\lambda}$ *is irreducible,*
2. $\rho_{f,\lambda}$ *is unramified outside Nl,*
3. *if $(p, Nl) = 1$ then* $\mathrm{tr}(\rho_{f,\lambda}(Frob_p)) = a_p$ *and* $\det(\rho_{f,\lambda}(Frob_p)) = \varepsilon(p)\, p^{k-1}$.

Remarks.

1. The representation $\rho_{f,\lambda}$ is uniquely determined by the given properties 1.-3. Indeed, by irreducibility it is uniquely determined by the characteristic polynomials and by Prop. 2.1 it suffices to look at Frobenius elements.
2. The representation $\rho_{f,\lambda}$ is constructed via the natural Galois representation on the étale l-adic cohomology of certain schemes over $\mathbb{Q}$ base extended to $\bar{\mathbb{Q}}$. We will not present this construction here but refer to [2] for all details.

To conclude these notes we briefly discuss the theorem in cases of low weight.

2.2.1 *Weight* 1

Let $k = 1$. In this case, $\rho_{f,\lambda}$ is actually defined over K and, extending linearly to $\mathbb{C}$, we obtain a continuous representation

$$\rho_f : G_{\mathbb{Q}} \longrightarrow GL_2(\mathbb{C}).$$

It has the properties:

1. ρ_f is irreducible with finite image,
2. ρ_f is unramified outside N,
3. if $(p, N) = 1$ then $\mathrm{tr}(\rho_f(Frob_p)) = a_p$ and $\det(\rho_f(Frob_p)) = \varepsilon(p)$.

By Prop. 2.1 the formula for det implies that $\det \rho_f = \varepsilon$ whence ρ_f is *odd*. Since the image is finite ρ_f induces an *Artin representation* of $Gal(L/\mathbb{Q})$ where $Gal(\bar{\mathbb{Q}}/L) := \ker \rho_f$. The following theorem is due to Deligne-Serre (cf. [3], Thm. 4.6).

Theorem 2.3. *The Artin conductor of ρ_f equals N and the Artin L-series $L(s, \rho_f)$ equals $L(s, f)$.*

Recall that the famous *Artin conjecture* says that $L(s, \rho)$ is entire for any irreducible Artin representation $\rho \neq 1$.

Corollary 2.4. *The Artin conjecture holds for* ρ_f.

Proof. By Thm. 1.14 the L-series $L(s, f)$ is entire.

2.2.2 *Weight* 2

Let $k = 2$. In this case, the image of $\rho_{f,\lambda}$ inside $GL_2(K_\lambda)$ is a compact subgroup whence an l-adic Lie group. It is not finite. Assume for simplicity $\varepsilon = 1$. To avoid a greater diversion into the theory of abelian varieties we assume that the level satisfies

$$N \in \{11, 14, 15, 17, 19, 20, 21, 24, 27, 32, 36, 49\}.$$

In this case the modular curve $X_0(N)$ corresponds via the usual GAGA-theorem to an elliptic curve over $\mathbb{Q}$. The space $S_2(N)$ can be identified with the space of holomorphic differentials on $X_0(N)$ of degree 1 whence $\dim_{\mathbb{C}} S_2(N) = 1$. Hence, there exists a *unique* primitive form $f \in S_2(N)$. It follows that $\rho_{f,\lambda} = \pi_l$ where

$$\pi_l : G_{\mathbb{Q}} \longrightarrow Aut_{\mathbb{Q}_l}(V_l(X_0(N)))$$

as in our previous example ($K = \mathbb{Q}$, $\lambda = l$).

For general N one has to pass to the Jacobian variety of $X_0(N)$ to realize the representation $\rho_{f,l}$.

References

[1] A. O. L. Atkin and J. Lehner, Hecke operators on $\Gamma_0(m)$, *Math. Ann.*, **185** (1970) 134–160.
[2] P. Deligne, Formes modulaires et représentations l-adiques (Sém. Bourb., vol. 1968/69, exp. 355). Lect. Notes in Math., Vol. 179. Springer-Verlag, Berlin (1971).
[3] P. Deligne and J.-P. Serre, Formes modulaires de poids 1, *Ann. Sci. École Norm. Sup. (4)*, **7** (1974) 507–530.
[4] F. Diamond and J. Shurman, A first course in modular forms, volume 228 of Graduate Texts in Math. Springer-Verlag, New York (2005).
[5] H. Hida, Geometric modular forms and elliptic curves. World Sci. Publ. Co. Inc., River Edge, NJ (2000).
[6] S. Lang, Introduction to modular forms, Springer-Verlag, Berlin (1976), Grundlehren der math. Wiss., No. 222.
[7] T. Miyake, Modular forms, Springer-Verlag, Berlin (1989).
[8] J.-P. Serre, Abelian l-adic representations and elliptic curves, volume 7 of Research Notes in Math. A K Peters Ltd., Wellesley, MA (1998).

Guwahati Workshop on Iwasawa Theory of Totally Real Fields No. 12, 2010, pp. 15–47.

Λ-adic Forms and the Iwasawa Main Conjecture

D. Banerjee, E. Ghate and V. G. Narasimha Kumar

School of Mathematics, Tata Institute of Fundamental Research, Homi Bhabha Road, Mumbai 400 005, India
e-mail: debargha@math.tifr.res.in; eghate@math.tifr.res.in; ganesh@math.tifr.res.in

1. Introduction

The classical Iwasawa main conjecture identifies two seemingly rather different power series in one variable over a p-adic integer ring, up to multiplication by a unit power series. One is the characteristic power series of a Selmer group, the other is a p-adic L-function. The conjecture over $\mathbb{Q}$ was proved by Mazur and Wiles, and over totally real fields by Wiles in [Wil90].

The purpose of these notes is to provide some background on Λ-adic forms to the extent they are used in Wiles' approach in [Wil90] to the proof of the main conjecture, assuming that the totally real base field is $\mathbb{Q}$. We first describe some general results about Λ-adic forms and their Galois representations, and then give an exposition of some specific results about Λ-adic forms in Sections 3 and 4 of [Wil90].

We leave it to the other articles in this volume (especially [CS10]) to explain how the material in these notes is used in the proof of the main conjecture (over $\mathbb{Q}$). However, some brief remarks on how Λ-adic forms enter the proof might be in order. To prove the main conjecture, it suffices to show that the two power series mentioned above have the same zeros, with multiplicity. Iwasawa had already shown that the sum of the multiplicities of the zeros is the same. Thus it suffices to show that the multiplicity of a specific zero of one of the power series (in practice one takes the characteristic power series of the Selmer group) is greater than the corresponding multiplicity of the corresponding zero of the other (the p-adic L-function). This is done by Lambda-fying, so to speak, the classical argument of Ribet [Rib76] (see also [Dal10]), proving the converse of Herbrand's theorem, and goes roughly as follows. The p-adic L-function turns out to be the constant term of a Λ-adic Eisenstein series (cf. Section 4). Thus, if a zero occurs in the p-adic L-function, then, going modulo the corresponding divisor, one expects a congruence between the Λ-adic Eisenstein series and a Λ-adic cuspidal eigenform F. The study of such congruences is made more precise by introducing the Λ-adic Eisenstein ideal (described in Section 8). One now uses the global Galois representation attached to the Λ-adic cusp form F, constructed in considerable detail in Section 6 (see especially Theorem 6.1), along with some important local properties of this representation (Theorem 6.2), to show that one can construct enough classes in the Selmer group so that a zero of greater multiplicity occurs in its characteristic power series. Thus, Λ-adic forms and their Galois representations are used in a rather fundamental way in Wiles' approach [Wil90] to the proof of the main conjecture.

The material in these notes is taken mostly from the papers of Hida and Wiles. For the background on Λ-adic forms and representations (described in Sections 2 through 6), we depend heavily on [Hid93, Chapter 7] and to some extent on [Wil88] and [Hid86]. The material in Sections 7 and 8 is taken almost verbatim from Sections 3 and 4 of [Wil90], though we provide detailed proofs of some of the results.

2. Some notation

We start by recalling some notation from Iwasawa theory. Let p be a prime and let $\mathbb{Z}_p$ denote the ring of integers of $\mathbb{Q}_p$. If $p = 2$, set $q = 4$, and if $p > 2$ set $q = p$. Let $W = 1 + q\mathbb{Z}_p$. Then the units of $\mathbb{Z}_p^\times$ decompose as:

$$\begin{aligned} \mathbb{Z}_p^\times &= (\mathbb{Z}/q)^\times \times W \\ x &= \omega(x) \times \langle x \rangle, \end{aligned}$$

where ω and $\langle\ \rangle$ are defined by the decomposition above.

Let u denote a topological generator of W. Let log be the usual p-adic logarithm from W to $q\mathbb{Z}_p$.

Lemma 2.1. *If* $z \in W = 1 + q\mathbb{Z}_p$, *then* $z = u^{s(z)}$, *where* $s(z) = \frac{\log(z)}{\log(u)} \in \mathbb{Z}_p$.

Proof. Let exp be the p-adic exponential map. Then

$$u^{s(z)} = \exp(s(z)\log(u)) = \exp\left(\frac{\log z}{\log u} \cdot \log u\right) = \exp(\log z) = z.$$

Here we are using the fact that exp induces an isomorphism between $q\mathbb{Z}_p$ and $1+q\mathbb{Z}_p$ with inverse log. □

Let N be an integer prime to p. It will denote tame level. We introduce three Dirichlet characters that occur throughout the exposition.

- χ will denote a fixed Dirichlet character of level Nq.
- ω above may be viewed as a Dirichlet character of conductor q. When p is odd, ω is just the mod p cyclotomic character. If $p = 2$, then ω is the non-trivial character of conductor 4.
- χ_ζ is a Dirichlet character of conductor p^r associated to a p-power root of unity as follows. If ζ has exact order p^{r-1} with $r \geq 1$ if p is odd, or exact order p^{r-2} with $r \geq 2$ if $p = 2$, define χ_ζ by mapping the image of $u \in W = 1 + q\mathbb{Z}_p$ in $(\mathbb{Z}/p^r)^\times$ to ζ.

Finally let ν_p denote the p-adic cyclotomic character, defined by $\zeta^g = \zeta^{\nu_p(g)}$, for all g in $\mathrm{Gal}(\bar{\mathbb{Q}}/\mathbb{Q})$ and all p-power roots of unity ζ.

Let $\Lambda = \mathbb{Z}_p[[X]]$ be the power series ring in one variable X over $\mathbb{Z}_p$. This is the usual Iwasawa algebra. If $s \in \mathbb{Z}_p$, then the coefficients of the power series

$$(1+X)^s = \sum_{m=0}^{\infty} \binom{s}{m} X^m$$

belongs to $\mathbb{Z}_p$, i.e., $(1+X)^s \in \Lambda$. Indeed if $P(s) \in \mathbb{Q}_p[s]$ is a polynomial in s, then it is easy to check that $P(s)$ is a continuous function from $\mathbb{Z}_p$ to $\mathbb{Q}_p$. In particular, the map $s \mapsto \binom{s}{m}$ is a continuous function from $\mathbb{Z}_p$ to $\mathbb{Q}_p$. This map takes $\mathbb{N}$ to $\mathbb{N}$. But $\mathbb{N}$ is dense in $\mathbb{Z}_p$. So this map induces a continuous function from $\mathbb{Z}_p$ to $\mathbb{Z}_p$. In particular, $\left|\binom{s}{m}\right|_p \leq 1$ if $s \in \mathbb{Z}_p$.

Let $\kappa : W = 1 + q\mathbb{Z}_p \to \Lambda^\times$ be the character which maps u to $1+X$, i.e., for each $s \in \mathbb{Z}_p$,

$$\kappa(u^s) = (1+X)^s = \sum_{m=0}^{\infty} \binom{s}{m} X^m \in \Lambda^\times.$$

The character κ may also be viewed as a Galois character via the natural map $\mathrm{Gal}(\bar{\mathbb{Q}}/\mathbb{Q}) \twoheadrightarrow \mathrm{Gal}(\mathbb{Q}_\infty/\mathbb{Q}) = W$, where $\mathbb{Q}_\infty$ is the cyclotomic $\mathbb{Z}_p$-extension of $\mathbb{Q}$. The character κ plays a central role in the subject. It is the universal deformation of the trivial mod p character; roughly it plays the role of the Λ-adic analogue of the cyclotomic character ν_p.

3. Λ-adic forms

Let p be an odd prime and let N be an integer prime to p. Let K be a finite field extension of the quotient field of Λ, and let I denote the integral closure of Λ in K. Observe that I is a complete local noetherian ring and is an integrally closed domain of Krull dimension 2. We remark that since I is a normal ring of dimension 2, it is Cohen-Macaulay [Mat86, Exer. 17.3], but unlike Λ, I is not necessarily regular.

Definition 3.1. *A Λ-adic form F of level N and character $\chi : (\mathbb{Z}/Np)^\times \to \mathbb{C}^\times$ is a formal q-expansion*

$$F = \sum_{n=0}^{\infty} a_n(F) q^n \in I[[q]]$$

such that for all specializations $\nu : I \to \bar{\mathbb{Q}}_p$, extending the usual specializations

$$\nu_{k,\zeta} : \Lambda \to \bar{\mathbb{Q}}_p$$
$$X \mapsto \zeta u^k - 1,$$

where $k > 1$ and $\zeta \in \mu_{p^{r-1}}$ with $r \geq 1$, the specialized q-expansion

$$f_\nu = \nu(F) = \sum_{n=0}^{\infty} \nu(a_n(F)) q^n \in \bar{\mathbb{Q}}_p[[q]]$$

is the image under a fixed embedding $\bar{\mathbb{Q}} \hookrightarrow \bar{\mathbb{Q}}_p$ of the q-expansion in $\bar{\mathbb{Q}}[[q]]$ of a classical modular form of

- *weight k,*
- *level Np^r, and,*
- *character $\chi_\nu = \chi\omega^{-k}\chi_\zeta$.*

Note that the character χ_ν has level Np^r with $r \geq 1$.

Thus a Λ-adic form is a family of classical forms of varying weights and level divisible by Np, with identical residual q-expansions. Explicit examples of Λ-adic forms are CM families (see, for instance, [Gha05, Sec. 5]) and Eisenstein families (described in Section 4).

Remark 3.2. Λ-adic forms may be defined for $p = 2$ as well. The definition is similar to the one given above for odd primes p, with some minor changes: one needs to replace p by $q = 4$ in the levels of χ and ω, and now $\zeta \in \mu_{p^{r-2}}$, $r \geq 2$, so that χ_ζ has level divisible by q. In particular, χ_ν has level Np^r with $r \geq 2$. We will treat the case $p = 2$ in the next section, but will later avoid $p = 2$ in these notes. We also remark that the symbol 'q' has two meanings: it is an odd prime or 4, and also $e^{2\pi i z}$ in various q-expansions. Since both usages are standard, we let this be; in any case which meaning is implied should be clear from the context.

Remark 3.3. Wiles uses a slightly different normalization in his papers [Wil88], [Wil90] when he defines a Λ-adic form. His specialization map $\nu_{k,\zeta}$ maps X to $\zeta u^{k-2} - 1$, rather than $\zeta u^k - 1$, and his f_ν is a weight k, level Np^r form, as above, but with character $\chi_\nu = \chi\omega^{2-k}\chi_\zeta$. Further, some authors (cf. Skinner's Hangzhou lectures notes [Ski04]) specialize X to $\zeta u^{k-1} - 1$ (with χ_ν now $\chi\omega^{1-k}\chi_\zeta$). We point out that all the above normalizations are in fact equivalent and one may go between them by a suitable automorphism of Λ. Indeed, we have the useful fact [Hid93, p. 199]:

Lemma 3.4. *The map induced by $X \mapsto aX + b$, where $a, b \in \mathbb{Z}_p$, and $|a|_p = 1$ and $|b|_p < 1$, is an automorphism of Λ.*

The different normalizations explain the occasional appearance of the terms u^2 (in Wiles' papers) or u (in Skinner's notes) in some of the formulas. As an aside, one may ask which normalization is best. It appears that the choice of $k - 2$ is historical and was used originally by Hida and then by Wiles, whereas the choice of k seems easiest for bookkeeping purposes, and is what we shall mainly try and use in these notes. However, we shall eventually switch to $k-2$ when exposing Wiles' results on Λ-adic forms from [Wil90, Sec. 4] in Section 8, to avoid introducing errors when changing normalizations. We also remark that some formulas, such as those for the Eisenstein family in the next section, or for the determinants of Λ-adic Galois representations, seem to be the most natural when one uses $k-1$, but we will *not* use this normalization in these notes.

Remark 3.5. There is no condition on the weight 1 specializations in the definition of a Λ-adic form, and the study of such specializations is an interesting area of research (cf. [MW86], [GV04]).

A Λ-adic form with q-expansion in $I[[q]]$ is sometimes more precisely referred to as an I-adic form. We shall use both terminologies. Let us write $M(N, \chi, I)$ for the I-module of I-adic forms of tame level N and character χ. Let

$$M(N, I) = \bigoplus_{\chi} M(N, \chi, I)$$

be the I-module of all I-adic forms (of some character χ).

Definition 3.6. *A Λ-adic form F is cuspidal if all the specializations f_ν above are cusp forms.*

We have the corresponding decomposition:

$$S(N, I) = \bigoplus_{\chi} S(N, \chi, I)$$

of the I-module of I-adic cusp forms.

4. Families of Eisenstein series

The aim of this section is to give an example of a Λ-adic form that is used extensively in Wiles' proof of the Iwasawa main conjecture. This form interpolates a family of Eisenstein series.

4.1 *Classical Eisenstein series*

Recall that if k is an even integer greater than or equal to 4, then the (normalized) Eisenstein series of weight k, level 1 and trivial character has q-expansion given by

$$E_k = \frac{\zeta(1-k)}{2} + \sum_{n=1}^{\infty} \sigma_{k-1}(n) q^n,$$

where $\sigma_{k-1}(n) = \sum_{d|n} d^{k-1}$ is the usual arithmetic function, and $\zeta(s)$ is the Riemann zeta function.

Let p be a prime. Since the forms E_k are of level 1 they cannot lie in a p-adic family (as we have seen forms f_ν in a p-adic family have level at least p, not to mention the fact that the character χ_ν depends on k, and cannot be uniformly trivial). Also, the Eisenstein series E_k of weight 2 and level 1 is not even holomorphic. However, for all even $k \geq 2$, the p-stabilized form $E_k^{(p)} = E_k(z) - p^{k-1} E_k(pz)$ is holomorphic and is of level p.

More generally, for any $k \geq 1$ and character ψ mod Np^r, with ψ having the same parity as k, consider the Eisenstein series of weight k, level equal to cond(ψ), and character ψ, given by:

$$E_{k,\psi} = \frac{L(1-k, \psi)}{2} + \sum_{n=1}^{\infty} \sigma_{k-1,\psi}(n) q^n,$$

where $\sigma_{k-1,\psi}(n) = \sum_{d|n} \psi(d)d^{k-1}$ and $L(s, \psi)$ is the Dirichlet L-series attached to ψ. Again, if ψ has level N (i.e., it has trivial p-part), then $E_{k,\psi}$ has level N, which is not divisible by Np. But, its p-stabilization, namely:

$$E^{(p)}_{k,\psi} = E_{k,\psi}(z) - \psi(p)p^{k-1}E_{k,\psi}(pz),$$

which has q-expansion:

$$E^{(p)}_{k,\psi} = \frac{L^{(p)}(1-k, \psi)}{2} + \sum_{n=1}^{\infty} \sigma^{(p)}_{k-1,\psi}(n)q^n,$$

where

$$\sigma^{(p)}_{k-1,\psi}(n) = \sum_{\substack{d|n \\ (d,p)=1}} \psi(d)d^{k-1},$$

and

$$L^{(p)}(s, \psi) = (1 - \psi(p)p^{-s})L(s, \psi)$$

is the Dirichlet L-series attached to ψ deprived of the Euler factor at p, has level divisible by Np. NB: $E^{(p)}_{k,\psi} = E_{k,\psi}$, if ψ already has conductor divisible by p. Given these remarks it seems that one should try and interpolate p-stabilized Eisenstein series. Before we do this we recall some classical facts about p-adic L-functions for $\mathrm{GL}_{1/\mathbb{Q}}$, which will be needed to interpolate the constant terms.

4.2 *Kubota-Leopoldt p-adic L-function*

Let χ be an arbitrary even Dirichlet character. The Kubota-Leopoldt p-adic L-function $L_p(s, \chi)$ attached to χ is a continuous function for $s \in \mathbb{Z}_p\backslash\{1\}$ (also continuous at $s = 1$, if χ is non-trivial), which satisfies the interpolation property

$$L_p(1-k, \chi) = (1 - \chi\omega^{-k}(p)p^{k-1})L(1-k, \chi\omega^{-k}), \tag{4.1}$$

for $k \geq 1$. Say χ is of type W if χ factors through W (that is, as a Galois character, χ factors through the Galois group $\mathrm{Gal}(\mathbb{Q}_\infty/\mathbb{Q})$, where $\mathbb{Q}_\infty$ is the cyclotomic $\mathbb{Z}_p$-extension of $\mathbb{Q}$). NB: the trivial character is of type W. Set

$$H_\chi(X) = \begin{cases} \chi(u)(1+X) - 1 & \text{if } \chi \text{ is of type } W, \\ 1 & \text{otherwise.} \end{cases}$$

Iwasawa showed [Was96, Thm. 7.10] that there exists a unique power series $G_\chi(X) \in I_\chi := \mathbb{Z}_p[\chi][[X]]$ such that

$$L_p(1-s, \chi) = \frac{G_\chi(u^s - 1)}{H_\chi(u^s - 1)}. \tag{4.2}$$

Moreover, if ρ is a character of type W, then (cf. [Wil90, (1.4)]):

$$G_{\chi\rho}(X) = G(\rho(u)(1+X) - 1). \tag{4.3}$$

This can be proved using [Was96, Thm. 7.10]; see also [Ven10, Prop. 2.1].

4.3 Λ-adic Eisenstein series

We are now ready to begin interpolating p-stabilized Eisenstein series. Recall that

- $q = p$ if p is odd, and,
- $q = 4$ if $p = 2$.

Fix an even character χ of conductor Nq. This will be the character of the Λ-adic form we construct (in the sense of Definition 3.1). For each $k > 1$ and

- $\zeta \in \mu_{p^{r-1}}$, with $r \geq 1$, if p is odd, and,
- $\zeta \in \mu_{p^{r-2}}$, with $r \geq 2$ if $p = 2$,

let $\psi = \chi_\nu = \chi\omega^{-k}\chi_\zeta$ be the character of level Np^r, which is the character of the form f_ν (see Definition 3.1). NB: Since χ is even, ψ has the same parity as k. Consider the corresponding Eisenstein series $E_{k,\psi}^{(p)}$ attached to $\psi = \chi\omega^{-k}\chi_\zeta$. This is a modular form in $M_k(Np^r, \psi)$.

Proposition 4.4. *Set $I_\chi = \mathcal{O}[[X]]$ with $\mathcal{O} = \mathbb{Z}_p[\chi]$. If $\chi \neq 1$, then there is a Λ-adic form*

$$\mathcal{E}_\chi = \sum_{n=0}^{\infty} A_{n,\chi}(X)q^n \in I_\chi[[q]],$$

which specializes to $E_{k,\psi}^{(p)}$, with $\psi = \chi\omega^{-k}\chi_\zeta$, under the homomorphism of I_χ to $\bar{\mathbb{Q}}_p$ induced by $\nu_{k,\zeta}$, for $k > 1$, ζ as above. If $\chi = 1$, then $\mathcal{E}_\chi$ still exists, but it is strictly speaking not a Λ-adic form, since the constant term of $\mathcal{E}_\chi$ has denominator X.

Proof. We first interpolate the non-constant terms. As remarked earlier, if $s \in \mathbb{Z}_p$, then

$$(1+X)^s = \sum_{m=0}^{\infty} \binom{s}{m} X^m \in \Lambda^\times.$$

By Lemma 2.1, if d is an integer with $d \equiv 1 \bmod q$, then $d = u^{s(d)}$ for $s(d) \in \mathbb{Z}_p$. Hence, if we set

$$A_d(X) = \frac{1}{d}(1+X)^{s(d)} \in \Lambda,$$

then

$$A_d(u^k - 1) = \frac{u^{s(d)k}}{d} = d^{k-1}.$$

This basic computation hints at what we should do for general d (coprime to p). Recall $\mathbb{Z}_p^\times = (\mathbb{Z}/q)^\times \times W$ with $x = \omega(x) \cdot \langle x \rangle$. Given d with $(d, p) = 1$, we have $\langle d \rangle \in W$. So define

$$A_d(X) = \frac{1}{d}(1+X)^{s(\langle d \rangle)}. \tag{4.5}$$

Then

$$A_d(\zeta u^k - 1) = \frac{\zeta^{s(\langle d \rangle)} u^{ks(\langle d \rangle)}}{d} = \frac{\chi_\zeta(\langle d \rangle)\langle d \rangle^k}{d} = \omega^{-k}(d)\chi_\zeta(d)d^{k-1}.$$

Now introduce the character χ of level Nq. For $n \geq 1$, set

$$A_{n,\chi}(X) = \sum_{\substack{d|n \\ (d,p)=1}} \chi(d) A_d(X).$$

Then

$$A_{n,\chi}(\zeta u^k - 1) = \sum_{\substack{d|n \\ (d,p)=1}} \psi(d) d^{k-1} = \sigma^{(p)}_{k-1,\psi}(n),$$

where $\psi = \chi\omega^{-k}\chi_\zeta$. Thus we have successfully interpolated the non-constant terms of $E^{(p)}_{k,\psi}$.

Now we need to interpolate the constant terms, that is, we need to find $A_{0,\chi}(X) \in I_\chi$ such that

$$A_{0,\chi}(\zeta u^k - 1) = \frac{L^{(p)}(1-k,\psi)}{2}.$$

But such a power series is furnished by the Kubota-Leopoldt p-adic L-function attached to χ. We define:

$$A_{0,\chi}(X) = \frac{1}{2}\frac{G_\chi(X)}{H_\chi(X)}.$$

Recall that in the present setting our χ has conductor Nq. If either the N-part or the p-part of χ is non-trivial, then χ is not of type W, and so if χ is non-trivial then $H_\chi(X) = 1$. If χ is trivial, then it is trivially of type W and $H_\chi(X) = X$. Thus $A_{0,\chi}(X) \in I_\chi$ if $\chi \neq 1$, and $XA_{0,\chi}(X) \in I_\chi$ if $\chi = 1$. Now:

$$\begin{aligned}
A_{0,\chi}(\zeta u^k - 1) &= \frac{1}{2}\frac{G_\chi(\zeta u^k - 1)}{H_\chi(\zeta u^k - 1)} \\
&= \frac{1}{2}\frac{G_{\chi\chi_\zeta}(u^k - 1)}{H_{\chi\chi_\zeta}(u^k - 1)} \qquad \text{by (4.3)} \\
&= \frac{L_p(1-k,\chi\chi_\zeta)}{2} \qquad \text{by (4.2)} \\
&= \frac{(1-(\chi\omega^{-k}\chi_\zeta)(p)p^{k-1})\cdot L(1-k,\chi\omega^{-k}\chi_\zeta)}{2} \qquad \text{by (4.1)} \\
&= \frac{L^{(p)}(1-k,\psi)}{2},
\end{aligned} \tag{4.6}$$

as desired.

Thus, if we define $\mathcal{E}_\chi = \sum_{n=0}^\infty A_{n,\chi}(X)q^n$, then $\mathcal{E}_\chi \in I_\chi[[q]]$ if $\chi \neq 1$ (and $X\mathcal{E}_\chi \in I_\chi[[q]]$ if $\chi = 1$), and $\mathcal{E}_\chi$ has the desired interpolation property. □

4.4 *Cuspidal families*

In this section we construct a cuspidal Λ-adic form starting with a classical cusp form of weight 1, and the Λ-adic form $\mathcal{E}_\chi$ just constructed. The argument is somewhat artificial so we will be brief.

Let $f_1 \in S_1(Nq, \psi_1)$ be a fixed cusp form of weight 1. Recall

$$\mathcal{E}_\chi(\zeta u^{k-1} - 1) = E^{(p)}_{k-1,\psi} \in M_{k-1}(Np^r, \psi)$$

where $\psi = \chi\omega^{1-k}\chi_\zeta$. Thus the product $f_1 \cdot E^{(p)}_{k-1,\psi}$ is a cusp form:

$$f_1 \cdot E^{(p)}_{k-1,\psi} \in S_k(Np^r, \chi'\omega^{-k}\chi_\zeta),$$

with $\chi' := \psi_1\chi\omega$. We show $f_1 \cdot E^{(p)}_{k-1,\psi}$ are the specializations at $X = \zeta u^k - 1$ of a cuspidal Λ-adic form F of level N and character χ' for $k > 1$ and ζ as above.

Assume that the q-expansions of f_1 and ψ_1 are both $\mathbb{Z}_p[\chi]$-rational (otherwise extend scalars). Then we may formally multiply the q-expansions in $I_\chi = \mathbb{Z}_p[\chi][[X]]$ of f_1 and $\mathcal{E}_\chi$. Say the resulting q-expansion is $f_1 \cdot \mathcal{E}_\chi = \sum_{n=0}^{\infty} a_n(X)q^n$, for some $a_n(X) \in I_\chi$. Now define

$$F := \sum_{n=0}^{\infty} a_n(u^{-1}X + u^{-1} - 1)q^n,$$

noting that the substitution made above is an automorphism of I_χ (cf. Lemma 3.4). Then on substituting $X = \zeta u^k - 1$ we get

$$F(\zeta u^k - 1) = \sum_{n=0}^{\infty} a_n(\zeta u^{k-1} - 1)q^n = f_1 \cdot \mathcal{E}_\chi(\zeta u^{k-1} - 1) = f_1 \cdot E^{(p)}_{k-1,\chi\omega^{1-k}\chi_\zeta}.$$

So F is the desired cuspidal family.

4.5 *Wiles' normalizations*

As mentioned in Remark 3.3, Wiles' normalization involves substituting $X = \zeta u^{k-2} - 1$, rather than $X = \zeta u^k - 1$. This changes the definition of $\mathcal{E}_\chi$ slightly. Since we will also work with Wiles' $\mathcal{E}_\chi$ below let us point out the changes that need to be made in the old definition:

(1) The denominator of d in (4.5) gets shifted to the numerator (i.e., multiply the old $A_d(X)$ by d^2) so that now

$$A_d(X) = d(1 + X)^{s(\langle d \rangle)},$$

and $A_{n,\chi}(X)$ for $n \geq 1$ is now defined with this new $A_d(X)$. We remark here that with 'Skinner's normalization' $X = \zeta u^{k-1} - 1$ one could simply drop the d altogether from $A_d(X)$, and so this normalization gives the neatest formula!.

(2) Let

$$\hat{G}_\chi(X) = G_{\chi\omega^2}(u^2(1 + X) - 1),$$
$$\hat{H}_\chi(X) = H_{\chi\omega^2}(u^2(1 + X) - 1).$$

Then the new constant term is defined as

$$A_{0,\chi}(X) = \frac{1}{2}\frac{\hat{G}_\chi(X)}{\hat{H}_\chi(X)}.$$

It is easily verified that the Wiles' Eisenstein family $\mathcal{E}_\chi = \sum_{n=0}^{\infty} A_{n,\chi}(X)q^n$ interpolates the p-stabilized Eisenstein series $E_{k,\psi}^{(p)}$ with $\psi = \chi\omega^{2-k}\chi_\zeta$ (note the $2-k$ instead of the $-k$) when one specializes at $X = \zeta u^{k-2} - 1$. This is more or less obvious for the non-constant terms; for the constant term we simply note:

$$\begin{aligned} A_{0,\chi}(\zeta u^{k-2} - 1) &= \frac{1}{2}\frac{\hat{G}_\chi(\zeta u^{k-2} - 1)}{\hat{H}_\chi(\zeta u^{k-2} - 1)} \\ &= \frac{1}{2}\frac{G_{\chi\omega^2}(\zeta u^k - 1)}{H_{\chi\omega^2}(\zeta u^k - 1)} \\ &= \frac{1}{2}\frac{G_{\chi\omega^2\chi_\zeta}(u^k - 1)}{H_{\chi\omega^2\chi_\zeta}(u^k - 1)} && \text{by (4.3)} \\ &= \frac{L^{(p)}(1-k, \chi\omega^{2-k}\chi_\zeta)}{2} && \text{as in (4.6).} \end{aligned}$$

5. Hida theory

In this section we mention some facts from Hida theory. From this point on, the prime p will be an odd prime. We start with some definitions.

Definition 5.1. *A Λ-adic form is a newform if each f_υ is N-new.*

Note that we do not require f_υ to be p-new; indeed the specializations f_ν may be p-old if $r = 1$.

The spaces $M(N, I)$ and $S(N, I)$ have a natural action of the Hecke operators T_q for $q \nmid Np$ and U_q for $q | Np$, given by the usual formulas. Sometimes we just write T_n for the n-th Hecke operator (including at the primes dividing Np). The Hecke action commutes with specialization.

Definition 5.2. *A Λ-adic form F is an eigenform if it is an eigenfunction of the Hecke operators. Equivalently, F is an eigenform if each specialization f_ν is an eigenform for the Hecke operators (in particular each f_ν is a U_p-eigenform).*

We now define the Λ-adic Hecke algebra $h(N, \Lambda)$ to be the Λ-subalgebra of the Λ-algebra $\mathrm{End}_\Lambda(M(N, \Lambda))$ generated by all the Hecke operators above. We define $h(N, I)$ by $h(N, \Lambda) \otimes_\Lambda I$. There is a bijection between I-algebra homomorphisms of $h(N, I)$ with values in I, and I-adic cuspidal Hecke eigenforms normalized to have first Fourier coefficient 1, induced by the map $(\lambda : h(N, I) \to I) \mapsto (F = \sum_{n=1}^{\infty} \lambda(T_n)q^n)$.

Definition 5.3. *A* Λ*-adic form is primitive if it is an eigenform, a newform, and normalized such that* $a_1(F) = 1$.

The spaces $M(N, I)$ and $S(N, I)$ of I-adic forms are not in general finitely generated as I-modules, even if one restricts to the new part. Put another way, the number of primitive I-adic forms of tame level N is not necessarily finite. Hida realized that just as holomorphic forms are more manageable in the class of real-analytic forms, in the p-adic world, ordinary forms are better behaved in the class of algebraic holomorphic forms.

Definition 5.4. *A* Λ*-adic form* F *is ordinary if* f_ν *is ordinary (that is,* $a_p(f_\nu)$ *is a* p*-adic unit), for all* ν *as above.*

Note the Eisenstein family $\mathcal{E}_\chi$ is ordinary, since $A_{p,\chi}(X) = 1$.

The sum of ordinary forms in the above sense (whether classical or Λ-adic) is not again necessarily ordinary in the above sense. To rectify this, Hida introduced a projector $e = \lim_{n\to\infty} U_p^{n!}$, and using e, defined the spaces of ordinary Λ-adic forms as follows:

$$M^{\mathrm{ord}}(N, I) = eM(N, I),$$

$$S^{\mathrm{ord}}(N, I) = eS(N, I).$$

Since $e(F) = F$ if F is ordinary in the sense of the previous definition, the sum of ordinary forms is now ordinary in the more general sense above. One of the key reasons the ordinary part is important is because of the following theorem.

Theorem 5.5. $M^{\mathrm{ord}}(N, I)$ *is finitely generated as an* I*-module.*

Proof. We prove this when $I = \Lambda$ since the proof is easily modified for general I. Note that after specialization at $\nu = \nu_{k,1}$, Λ-adic forms in $M^{\mathrm{ord}}(N, \Lambda)$ wind up in the classical space $M_k^{\mathrm{ord}}(\Gamma_1(Np), \mathbb{Z}_p) = eM_k(\Gamma_1(Np), \mathbb{Z}_p)$. The proof now involves two parts. One first shows that the spaces $M_k^{\mathrm{ord}}(\Gamma_1(Np), \mathbb{Z}_p)$ have rank independent of k. One then shows that this forces $M^{\mathrm{ord}}(N, \Lambda)$ to be finitely generated as a Λ-module.

Let us deal with the first part first. By duality it suffices to show that the Hecke algebra acting on $M_k^{\mathrm{ord}}(\Gamma_1(Np), \mathbb{Z}_p)$ has rank independent of k. By the Eichler-Shimura isomorphism it further suffices to show that $\mathrm{H}^1(\Gamma_1(Np), L(n, \mathbb{Z}_p))$ has rank independent of n, where $n = k - 2$, and $L(n, A) = \mathrm{Sym}^n(A^2)^*$ is the space of homogeneous polynomials in two variables with coefficients in A. But now a pretty cohomological argument (see [Hid93, p. 203]) shows that the rank is in fact bounded by dimension $e\mathrm{H}^1(\Gamma_1(Np), \mathbb{F}_p)$. We remark that since $\mathbb{F}_p = L(0, \mathbb{F}_p) = L(2-2, \mathbb{F}_p)$, the number of ordinary forms of all weights and level Np is controlled by the weight 2 ordinary forms.

Now say that M is a finitely generated free submodule of $M^{\mathrm{ord}}(N, \Lambda)$. Choose a basis $F_1, F_2, \ldots, F_d$. Choose integers $n_1, n_2, \ldots n_d$ such that $\det(a_{n_i}(F_j)) \neq 0$ in Λ. By the Weierstrass preparation theorem, there is a $k > 1$ such that the specialization of $\det(a_{n_i}(F_j))$ at $u^k - 1$ is also not 0. Let $f_1, f_2, \ldots, f_d$ be the corresponding specializations at $\nu_{k,1}$; by hypothesis they generate a free rank d submodule of

$M_k^{\mathrm{ord}}(\Gamma_1(Np), \mathbb{Z}_p)$. But as we have mentioned above this last module has bounded rank, and so d is bounded from above. In particular the maximal rank free submodule of $M^{\mathrm{ord}}(N, \Lambda)$ has bounded rank, say r.

Now we can show that $M^{\mathrm{ord}}(N, \Lambda)$ is finitely generated as a Λ-module. Suppose $F_1, F_2, \ldots, F_r$ is a maximal linearly independent set of elements in $M^{\mathrm{ord}}(N, \Lambda)$. Then these elements form a basis of $M^{\mathrm{ord}}(N, \Lambda) \otimes_\Lambda Q(\Lambda)$, where $Q(\Lambda)$ is the quotient field of Λ. Thus, every element of $F \in M^{\mathrm{ord}}(N, \Lambda)$ is a linear combination $F = \sum_{i=1}^r x_i F_i$ of these F_i's with coefficients $x_i \in Q(\Lambda)$. For any set of r natural numbers $n_1, n_2, \ldots, n_r$, we have the matrix equation

$$AX = B$$

where $A = (a_{n_i}(F_j))$, $X = (x_1, x_2, \ldots, x_r)^t$ and $B = (a_{n_1}(F), \ldots, a_{n_r}(F))^t$. Choose the n_i such that $D = \det(A) \in \Lambda$ is non-zero. Multiplying the above matrix equation by the adjoint matrix of A, we see that $Dx_i \in \Lambda$, for $i = 1, \ldots, r$. Hence $DM^{\mathrm{ord}}(N, \Lambda) \subseteq \Lambda F_1 + \cdots + \Lambda F_r$. It follows that $M^{\mathrm{ord}}(N, \Lambda) \cong DM^{\mathrm{ord}}(N, \Lambda)$ is finitely generated as a Λ-module, since is isomorphic to a submodule of a finitely generated Λ-module, and Λ is noetherian. □

Remark 5.6. Clearly since $M(N, \Lambda)$ is a submodule of $\Lambda[[q]]$ it is torsion-free. With a bit more effort one can show that the finitely generated torsion-free Λ-module is in fact Λ-free (see [Hid93, p. 209]).

We now define the ordinary Λ-adic Hecke algebra $h^{\mathrm{ord}}(N, \Lambda)$ to be the Λ-subalgebra of $\mathrm{End}_\Lambda(M^{\mathrm{ord}}(N, \Lambda))$ generated over Λ by all the Hecke operators above. Similarly we define $h^{\mathrm{ord}}(N, I)$ by $h^{\mathrm{ord}}(N, \Lambda) \otimes_\Lambda I$. The finiteness of the space of ordinary Λ-adic forms was initially proved by Hida who showed that $h^{\mathrm{ord}}(N, \Lambda)$ is finitely generated as a Λ-module. This statement already implies the first part of the following theorem which summarizes some of the main results of Hida theory.

Theorem 5.7 (Hida, Wiles). *Let p be an odd prime.*

(1) *There are finitely many primitive, ordinary Λ-adic forms F of tame level N.*
(2) *Each classical, p-stabilized, primitive, ordinary form lives in some primitive, ordinary form F.*
(3) *The form F in part* (2) *is unique up to Galois conjugacy.*
(4) *Given a normalized, ordinary, Λ-adic eigenform F, one may associate a Galois representation ρ_F to it, with several natural properties.*

Parts (2) and (3) follow from what is usually referred to as Hida's control theorem and are not proved in these notes. The representation ρ_F in part (4) was constructed by Hida. In the next few sections we explain an alternative method of Wiles for constructing ρ_F. We also describe the local behaviour of ρ_F (a result of Wiles). This last result plays a key role in the proof of the Iwasawa main conjecture.

6. Ordinary Λ-adic Galois representations

The goal of this section is to construct Galois representations attached to ordinary Λ-adic eigenforms. All the material here is either taken directly (or modified slightly) from [Hid93] or [Wil88].

In this section p is an odd prime. Recall Λ denotes the power series ring $\mathbb{Z}_p[[X]]$, K is a finite extension of the field of fractions of Λ, and I denotes the integral closure of Λ in K. We shall explain Wiles' method of attaching a Galois representation into $\mathrm{GL}_2(K)$ to a normalized ordinary I-adic cuspidal eigenform F.

6.1 *Results*

We first state the precise result.

Theorem 6.1 (Hida, Wiles). *Let F be a normalized I-adic eigenform of level N in $S^{\mathrm{ord}}(N, \chi, I)$, and let λ denote the corresponding I-algebra homomorphism $\lambda : h^{\mathrm{ord}}(N, \chi; I) \to I$. Then there exists a unique Galois representation $\rho_F : \mathrm{G}_{\mathbb{Q}} = \mathrm{Gal}(\bar{\mathbb{Q}}/\mathbb{Q}) \to \mathrm{GL}_2(K)$ such that*

(1) *ρ_F is continuous and absolutely irreducible,*
(2) *ρ_F is unramified outside Np,*
(3) *for each prime $q \nmid Np$, we have*

$$\det(1 - \rho_F(\mathrm{Frob}_q)T) = 1 - \lambda(T_q)T + (\chi\kappa\nu_p^{-1})(q)\, T^2,$$

where Frob_q is the Frobenius element at q and $\kappa : W = 1 + p\mathbb{Z}_p \to \Lambda^\times$ is the usual character, and for any $x \in \mathbb{Z}_p^*$, $\langle x \rangle := \omega(x)^{-1}x \in W$.

As mentioned above, the global representation ρ_F is absolutely irreducible. However the local representation obtained by restricting $\rho = \rho_F$ to a decomposition group D_p at the prime p, is reducible. More precisely, we have:

Theorem 6.2 (Wiles). *Maintaining the same notations as above, the restriction of ρ_F to D_p is given up to equivalence by*

$$\rho_F|_{D_p} \sim \begin{pmatrix} \epsilon_1 & * \\ 0 & \epsilon_2 \end{pmatrix},$$

where ϵ_2 is unramified and $\epsilon_2(\mathrm{Frob}_p) = \lambda(T_p)$.

We shall prove Theorem 6.1 over the next several sections. Before we start, let us explain the strategy of the proof. Starting with a family of representations $\rho_i : \mathrm{G}_{\mathbb{Q}} \to \mathrm{GL}_2(M_i)$, where M_i is a finite extension of $\mathbb{Q}_p$, we wish to construct a Λ-adic representation $\rho : \mathrm{G}_{\mathbb{Q}} \to \mathrm{GL}_2(K)$ such that the 'reduction' of the representation ρ (see §6.3 below) at certain height 1 prime ideals $\{P_i\}_{i=1}^{\infty}$ is isomorphic to ρ_i (over some finite extension of $\mathbb{Q}_p$). In our case, the representations $\{\rho_i\}_{i=1}^{\infty}$ come from the representations associated to the classical specializations $F(P_i)$ (see §6.5 below), where P_i's are height 1 primes that are the kernels of the specialization maps $\nu = \nu_i$ considered earlier; the resulting representation ρ will be ρ_F. In order to patch the representations $\{\rho_i\}_{i=1}^{\infty}$ into a Λ-adic representation, a new notion was introduced by Wiles in [Wil88], namely that of a pseudo-representation (see §6.6 below). The importance of this notion lies in the fact that it is easier to patch pseudo-representations than it is to patch usual representations.

6.2 *Continuity of the Galois representation*

Before we prove Theorem 6.1, let us comment on what it means for a Λ-adic Galois representation $\rho : \mathrm{G}_{\mathbb{Q}} \to \mathrm{GL}_2(K)$ to be continuous.

Definition 6.3. *A Galois representation $\rho : \mathrm{G}_{\mathbb{Q}} \to \mathrm{GL}_2(K)$ is said to be continuous if there is an I-submodule $\mathcal{L}$ of K^2 (called a lattice) such that $\mathcal{L}$ is of finite type over I, $\mathcal{L} \otimes_I K = K^2$, $\mathcal{L}$ is stable under ρ, and as a map $\rho : \mathrm{G}_{\mathbb{Q}} \to Aut_I(\mathcal{L})$, ρ is continuous, where $Aut_I(\mathcal{L})$ is equipped with the projective limit topology*

$$Aut_I(\mathcal{L}) = \varprojlim_i Aut_I(\mathcal{L}/\mathbf{m}^i \mathcal{L})$$

for the unique maximal ideal $\mathbf{m}$ of I.

Remark 6.4 ([Hid93, §7.5]). Since I is a ring of Krull dimension 2, its field of fractions K is not locally compact in any non-discrete topology on K (cf. Bourbaki, Commutative Algebra, §9.3). This implies that the image of a continuous representation of $\mathrm{G}_{\mathbb{Q}}$ into $\mathrm{GL}_2(K)$, with non-discrete topology on K, is small. This is why one takes the $\mathbf{m}$-adic topology on $\mathrm{End}_I(\mathcal{L})$ to define the continuity of ρ.

Remark 6.5. It is easy to see that an I-module $\mathcal{L}$ as above exists. However, $\mathcal{L}$ may not be free over I, and so one takes the projective limit topology on $\mathrm{Aut}_I(\mathcal{L})$. One thinks of the projective limit topology as a slight generalisation of the 'usual topology' on $\mathrm{GL}_2(I)$, which is induced from the product topology on I^4, since if $\mathcal{L}$ is a free module of rank 2 over I, then the projective limit topology on $\mathrm{Aut}_I(\mathcal{L})$ coincides with the 'usual topology'. Also, the continuity of the representation does not depend on the choice of the lattice $\mathcal{L}$, because of the Artin-Rees lemma, according to [Hid93, p. 228].

6.3 *Reduction of the representation ρ_F modulo a non-zero prime ideal P*

We wish to speak of the reduction of ρ_F modulo a non-zero prime ideal P, even before we have constructed the representation ρ_F. For any prime ideal P of I, let $Q(I/P)$ denote the field of fractions of I/P.

Definition 6.6. *A Galois representation $\rho_F(P)$ into $\mathrm{GL}_2(Q(I/P))$ is called a residual representation of ρ_F at P, if $\rho_F(P)$ is continuous under the $\mathbf{m}$-adic topology of $Q(I/P)$, it is semi-simple, and it satisfies the following properties:*

- *$\rho_F(P)$ is unramified outside Np,*
- *For any prime $q \nmid Np$,*

$$\det(1 - \rho_F(P)(\mathrm{Frob}_q)T) = 1 - \lambda(T_q)(P)T + ((\chi\kappa\nu_p^{-1})(q))(P)T^2.$$

6.4 *Existence of the residual representation at a non-zero prime P*

A priori it is not clear that the residual representation $\rho_F(P)$ attached to ρ_F exists. However, for any non-zero prime ideal, the residual representation exists.

Theorem 6.7. *For every prime ideal P of height* 1, *the residual representation $\rho_F(P)$ of ρ_F exists and it is unique up to an isomorphism over $Q(I/P)$.*

Proof. Let $\mathcal{L}$ be a lattice such that the image of ρ_F is a subset of $\mathrm{Aut}_I(\mathcal{L})$. We know that I is a noetherian integrally closed domain of dimension 2. Let P denote a prime ideal of height 1. Let I_P denote the localisation of I at P. It is easy to see that I_P is a DVR. Since I_P is a PID, $\mathcal{L}_P = \mathcal{L} \otimes_I I_P$ becomes a free module of rank 2 over I_P. Identifying $\mathcal{L}_P$ with I_P^2 we can view the representation ρ_F as $\rho_F : \mathrm{G}_{\mathbb{Q}} \to \mathrm{GL}_2(I_P)$. Reducing ρ_F modulo P, and noting $I_P/PI_P = Q(I/P)$, we get

$$\rho_P : \mathrm{G}_{\mathbb{Q}} \to \mathrm{GL}_2(Q(I/P)).$$

Let $\rho_F(P)$ be the semi-simplification of ρ_P. This is a candidate for the residual representation of ρ at P. Clearly, $\rho_F(P)$ satisfies the conditions mentioned in Definition 6.6, since ρ_F does. For example, the continuity of $\rho_F(P)$ follows from that of ρ_F, noting that after localizing $\mathcal{L}_P$ becomes free module over I_P, hence the projective limit topology and the usual topology coincide. The uniqueness of the representation $\rho_F(P)$ follows from fact that the semi-simplification of a representation over a field of characteristic 0 is completely determined by its traces (see Lemma 6.8 below). The same result shows that $\rho_F(P)$ is independent of the choice of lattice $\mathcal{L}$. □

Lemma 6.8 ([Ser89, Chap. 1, §2]). *Let G be a group, K be a field of characteristic* 0 *and let ρ_1 and ρ_2 be two finite dimensional linear representations of G over K. If ρ_1 and ρ_2 are semi-simple and* Trace $(\rho_1(g))$ = Trace $(\rho_2(g)), \forall\, g \in G$, *then ρ_1 and ρ_2 are isomorphic over K.*

Remark 6.9. By induction on the height of prime ideals, one can define the residual representation for prime ideals of height 2 as well. For further details, see [Hid93, §7.3].

6.5 *Identifying the residual representations $\rho_F(P)$ for arithmetic prime P*

We will in particular be interested in identifying the reductions of ρ_F at certain special height one prime ideals of I. Let

$$\Xi(I) = \{P : P = \text{kernel of a specialization map } \nu : I \to \bar{\mathbb{Q}}_p \text{ as in definition 3.1}\}.$$

The elements of $\Xi(I)$ are clearly height one prime ideals of I and are called arithmetic primes. We sometimes identify the homomorphisms ν with their kernels P.

Any element $\alpha \in I$ can be viewed as a function on $\mathrm{Spec}(I)(\bar{\mathbb{Q}}_p) = \mathrm{Hom}(I, \bar{\mathbb{Q}}_p)$ by $\alpha(P) := P(\alpha) = \alpha \pmod P$. In particular $F(P)$ gives a formal q-expansion in $\bar{\mathbb{Q}}_p[[q]]$. If P is an arithmetic prime, then $F(P)$ is by definition a classical cuspidal eigenform of weight at least 2, and so there is a classical Galois representation $\rho_{F(P)}$ attached to $F(P)$ by Deligne. We show that this representation agrees with the residual representation $\rho_F(P)$ attached to ρ_F.

First let us recall the definition of Galois representations attached to classical cusp forms. Let ψ be a character of conductor Np^r. Define

$$S_k(\Gamma_0(Np^r), \psi; A) := S_k(\Gamma_0(Np^r), \psi; \mathbb{Z}[\psi]) \otimes_{\mathbb{Z}[\psi]} A,$$

for any algebra A in $\mathbb{C}$ or $\bar{\mathbb{Q}}_p$, containing $\mathbb{Z}[\psi]$. Let $h_k(\Gamma_0(Np^r), \psi; A)$ denote the corresponding Hecke algebra. We have the following duality between cusp forms and the Hecke algebra:

$$\mathrm{Hom}_{\mathbb{Z}[\psi]}(h_k(\Gamma_0(Np^r), \psi; A), A) \xrightarrow{\sim} S_k(\Gamma_0(Np^r), \psi; A)$$

$$\varphi \mapsto \sum_{n=1}^{\infty} \varphi(T(n))q^n. \tag{6.10}$$

Under this duality normalized eigenforms correspond to $\mathbb{Z}[\psi]$-algebra homomorphisms from $h_k(\Gamma_0(Np^r), \psi; A)$ to A. Let f be a normalised eigenform in $S_k(\Gamma_0(Np^r), \psi; K_f)$ where K_f denotes the number field generated by $\{a_n(f)\}_{n=1}^{\infty}$. Let λ_f denote the corresponding algebra homomorphism $h_k(\Gamma_0(Np^r), \psi; K_f) \to K_f$. Then we have the following:

Theorem 6.11. *For each maximal ideal $\wp$ of $\mathcal{O}_{K_f}$ lying over p, there exists a unique 2-dimensional Galois representation $\rho_{f,\wp} : \mathrm{G}_{\mathbb{Q}} \to \mathrm{GL}_2(K_{f,\wp})$ such that*

(1) *$\rho_{f,\wp}$ is continuous and absolutely irreducible,*
(2) *$\rho_{f,\wp}$ is unramified outside Np,*
(3) *for each prime $q \nmid Np$, we have*

$$\det(1 - \rho_{f,\wp}(\mathrm{Frob}_q)T) = 1 - \lambda_f(T_q)T + \psi(q)q^{k-1}T^2.$$

The existence of such a representation was proved by Eichler-Shimura when $k = 2$, and by Deligne for $k \geq 2$.

As mentioned above, the representation $\rho_{f,\wp}$ is irreducible. However the local representation obtained by restricting $\rho_{f,\wp}$ to a decomposition subgroup D_p at the prime p, is reducible. More precisely, we have:

$$\rho_{f,\wp}|_{D_p} \sim \begin{pmatrix} \psi\nu_p^{k-1}\,\lambda(a_p)^{-1} & * \\ 0 & \lambda(a_p) \end{pmatrix}, \tag{6.12}$$

where $\lambda(a)$ is the unramified character of D_p taking $\mathrm{Frob}_p \in D_p/I_p$ to a and ν_p is the p-adic cyclotomic character.

Now let P be a prime corresponding to ν extending $\nu_{k,\zeta}$, i.e., an arithmetic prime. Let $f = f_\nu = F(P)$. Let $\wp$ be the prime of K_f induced by the fixed embedding $\bar{\mathbb{Q}} \hookrightarrow \bar{\mathbb{Q}}_p$. By the computations in (6.14) below, we see that $K_{f,\wp} \subseteq Q(I/P)$ and hence $\mathcal{O}_{K_{f,\wp}} \subseteq \widetilde{I/P}$, where $\widetilde{I/P}$ is the integral closure of I/P in $Q(I/P)$. We can state a refined version of Theorem 6.11, for $f = F(P)$, as follows:

Corollary 6.13. *There exists a unique, odd, Galois representation $\rho_{F(P)} : \mathrm{G}_{\mathbb{Q}} \to \mathrm{GL}_2(Q(I/P))$, and hence into $\mathrm{GL}_2(\widetilde{I/P})$, with the properties mentioned in Theorem* 6.11.

We can now show that the residual representation $\rho_F(P)$ attached to ρ_F is nothing but Deligne's representation $\rho_{F(P)}$, as Galois representation into $\mathrm{GL}_2(Q(I/P))$.

To do this it is enough to show that the characteristic polynomials of the Frobenius elements outside Np coincide (by Lemma 6.8, and noting that the Frobenius elements outside Np are dense in $G_{\mathbb{Q}}$). We compare the polynomials in Definition 6.6 and Theorem 6.11. Recall that for an integer n prime to p, we defined $s(\langle n\rangle) \in \mathbb{Z}_p$ so that $\langle n\rangle = u^{s(\langle n\rangle)}$. Now

$$\kappa(\langle n\rangle)(\zeta u^k - 1) = \kappa(u^{s(\langle n\rangle)})(\zeta u^k - 1) = (\zeta u^k)^{s(\langle n\rangle)} = \chi_\zeta(n)\omega^{-k}n^k,$$

since $\chi_\zeta(n) = \chi_\zeta(\langle n\rangle) = \zeta^{s(\langle n\rangle)}$ and $\langle n\rangle = \omega^{-1}(n)n$. Hence, if $P|_\Lambda = \ker(\nu_{k,\zeta})$, then

$$\lambda(T_q)(P) = a_q(F)(P) = a_q(F(P)) = a_q(f_\nu), \quad \text{and}$$

$$((\chi\kappa\nu_p^{-1})(q))(P) = (\chi\omega^{-k}\chi_\zeta)(q)q^{k-1} = \chi_\nu(q)q^{k-1}, \tag{6.14}$$

as desired.

6.6 *Pseudo-representations*

Let B be a commutative topological ring with unity and assume that 2 is invertible in B. Further, if B is an integral domain, let $Q(B)$ denote the field of fractions of B.

Definition 6.15. *Let G be a pro-finite group with an identity e, and a special element c of order 2. A pseudo-representation π from G to B, which shall be denoted by $\pi : G \to B$, is a triple $\pi = (A, D, X)$ of continuous maps*

$A : G \to B$,
$D : G \to B$, *and,*
$X : G \times G \to B$,

satisfying the following axioms:

(1) $A(\sigma\tau) = A(\sigma)A(\tau) + X(\sigma, \tau)$,
(2) $D(\sigma\tau) = D(\sigma)D(\tau) + X(\tau, \sigma)$,
(3) $X(\sigma\tau, \gamma) = A(\sigma)X(\tau, \gamma) + D(\tau)X(\sigma, \gamma)$,
(4) $X(\sigma, \tau\gamma) = A(\gamma)X(\sigma, \tau) + D(\tau)X(\sigma, \gamma)$,
(5) $A(e) = D(e) = A(c) = 1, D(c) = -1$,
(6) $X(\sigma, e) = X(e, \sigma) = X(\sigma, c) = X(\sigma, c) = 0$,
(7) $X(\sigma, \tau)X(\gamma, \eta) = X(\sigma, \eta)X(\gamma, \tau)$.

Remark 6.16. We denote A, D, X by A_π, D_π, X_π, when we need to specify π explicitly.

The function X is in fact determined by the function A, since for any $\sigma, \tau \in G$, we have $X(\sigma, \tau) = A(\sigma\tau) - A(\sigma)A(\tau)$. So it is possible to give an alternative description of a pseudo-representation using only the maps A and D. However, the given definition is convenient because it turns out to be important whether or not $X(\sigma, \tau) = 0$.

The name pseudo-representation implies, of course, a relationship with representations. Suppose ρ is an odd, 2-dimensional representation from G to B. By this we mean $\rho : G \to \mathrm{GL}_2(B)$ is a continuous homomorphism with $\rho(c) = \left(\begin{smallmatrix} 1 & 0 \\ 0 & -1 \end{smallmatrix}\right)$. Letting $\rho(\sigma) = \left(\begin{smallmatrix} a(\sigma) & b(\sigma) \\ c(\sigma) & d(\sigma) \end{smallmatrix}\right)$, one verifies immediately that $\pi = (A, D, X)$ given by

$$A(\sigma) = a(\sigma), \;\; D(\sigma) = d(\sigma), \;\; X(\sigma, \tau) = b(\sigma)c(\tau),$$

is a pseudo-representation from G to B.

Definition 6.17. *If $\pi : G \to B$ is as above, then we say that π is the pseudo-representation* associated *to ρ, or that π* comes from *ρ.*

Given a pseudo-representation π, we can define the Trace and Determinant of π.

Definition 6.18. $\mathrm{Tr}(\pi)(\sigma) := A(\sigma) + D(\sigma), \;\; \mathrm{Det}(\pi)(\sigma) := A(\sigma)D(\sigma) - X(\sigma, \sigma).$

Remark 6.19. It is clear that, if π comes from a representation, then the definitions of the Trace and Determinant of π coincide with that of the representation.

We have the following easily checked identities for $A(\sigma)$ and $D(\sigma)$, which play an important role in the proofs:

$$A(\sigma) = \frac{\mathrm{Tr}(\pi)(\sigma) + \mathrm{Tr}(\pi)(c\sigma)}{2}, \quad \text{and,}$$
$$D(\sigma) = \frac{\mathrm{Tr}(\pi)(\sigma) - \mathrm{Tr}(\pi)(c\sigma)}{2}. \tag{6.20}$$

Here is a natural question: When does a pseudo-representation come from a representation? This question has an affirmative answer in some cases, as we see now.

Theorem 6.21. *Let π be a pseudo-representation from G to B such that either X is identically zero or $X(h_1, h_2) \in B^*$ for some $h_1, h_2 \in G$. Then π comes from a 2-dimensional odd representation $\rho : G \to \mathrm{GL}_2(B)$.*

Proof. Let us divide the proof into two cases:

Case (i). If $X \equiv 0$, then we can define ρ by letting, for $g \in G$,

$$\rho(g) = \begin{pmatrix} A_\pi(g) & 0 \\ 0 & D_\pi(g) \end{pmatrix}.$$

Since X is identically zero, ρ is a representation from $G \to GL_2(B)$.

Case (ii). Suppose $\exists$ two elements $h_1, h_2 \in G$ such that $X(h_1, h_2) \in B^*$. We define the representation ρ by setting, for $g \in G$,

$$\rho(g) = \begin{pmatrix} a(g) & b(g) \\ c(g) & d(g) \end{pmatrix},$$

with $a(g) = A(g), d(g) = D(g), b(g) = \frac{X(g,h_2)}{X(h_1,h_2)}, c(g) = X(h_1, g)$. It is easy to check that ρ is a representation, by using the axioms for pseudo-representations (for the calculations see [Hid93, §7.5, Prop. 1]). □

Corollary 6.22. *If B is a field, then the pseudo-representation π always comes from a 2-dimensional, odd, representation.*

Let us return to our situation where $G = \mathrm{G}_{\mathbb{Q}}$ and I is a finite extension of Λ. As mentioned earlier, the proof of Theorem 6.1 depends on patching results for pseudo-representations, which we describe now.

Lemma 6.23. *Let $\mathfrak{a}$ and $\mathfrak{b}$ be ideals of I. Let $\pi(\mathfrak{a})$ and $\pi(\mathfrak{b})$ be pseudo-representations of G into $I/\mathfrak{a}$ and $I/\mathfrak{b}$, which are compatible, i.e., there exist two functions T and D on a dense subset Σ of G, with values in I, such that*

$$\mathrm{Tr}(\pi(\mathfrak{a}))(\sigma) \equiv T(\sigma) \pmod{\mathfrak{a}} \quad \text{and} \quad \mathrm{Tr}(\pi(\mathfrak{b}))(\sigma) \equiv T(\sigma) \pmod{\mathfrak{b}},$$
$$\mathrm{Det}(\pi(\mathfrak{a}))(\sigma) \equiv D(\sigma) \pmod{\mathfrak{a}} \quad \text{and} \quad \mathrm{Det}(\pi(\mathfrak{b}))(\sigma) \equiv D(\sigma) \pmod{\mathfrak{b}}, \tag{6.24}$$

for all $\sigma \in \Sigma$. Then, there exists a pseudo-representation$\pi(\mathfrak{a}\cap\mathfrak{b})$ of G into $I/(\mathfrak{a}\cap\mathfrak{b})$, such that

$$\mathrm{Tr}(\pi(\mathfrak{a}\cap\mathfrak{b}))(\sigma) \equiv T(\sigma) \pmod{\mathfrak{a}\cap\mathfrak{b}} \text{ and } \mathrm{Det}(\pi(\mathfrak{a}\cap\mathfrak{b}))(\sigma) \equiv D(\sigma) \pmod{\mathfrak{a}\cap\mathfrak{b}},$$

on Σ.

Proof. Let us briefly explain the proof. We have the following short exact sequence

$$0 \to I/(\mathfrak{a}\cap\mathfrak{b}) \to I/\mathfrak{a} \oplus I/\mathfrak{b} \to I/(\mathfrak{a}+\mathfrak{b}) \to 0$$
$$\bar{a} \mapsto (\bar{a}, \bar{a})$$
$$(\bar{a}, \bar{b}) \mapsto \overline{a-b}.$$

Consider the map $\pi = \pi(\mathfrak{a}) \oplus \pi(\mathfrak{b})$ from G to $I/\mathfrak{a} \oplus I/\mathfrak{b}$. Since, $\pi(\mathfrak{a})$ and $\pi(\mathfrak{b})$ are pseudo-representations, it is easy to see π is also a pseudo-representation into $I/\mathfrak{a} \oplus I/\mathfrak{b}$. By (6.24), $\mathrm{Tr}(\pi)$ mod $(\mathfrak{a}+\mathfrak{b})$ vanishes identically on Σ, and hence $\mathrm{Tr}(\pi)(\sigma) \equiv T(\sigma)$ mod $(\mathfrak{a}\cap\mathfrak{b})$, for all $\sigma \in \Sigma$. So π is a candidate for $\pi(\mathfrak{a}\cap\mathfrak{b})$. It suffices to show π takes values in $I/\mathfrak{a}\cap\mathfrak{b}$. Since $\mathrm{Tr}(\pi)$ is a continuous map, the zero ideal is closed in $I/(\mathfrak{a}+\mathfrak{b})$ and Σ is dense in G, $\mathrm{Tr}(\pi)$ mod $(\mathfrak{a}+\mathfrak{b})$ vanishes on G, and hence $\mathrm{Tr}(\pi)$ indeed takes values in $I/\mathfrak{a}\cap\mathfrak{b}$. Now, by (6.20), it is clear that the functions A_π, D_π, X_π corresponding to π, take values in $I/\mathfrak{a}\cap\mathfrak{b}$, as desired. □

We wish to generalize the lemma to a countable collection of ideals. Before stating this result, let us recall some results from commutative algebra.

Lemma 6.25. *Let A be a complete semi local ring, $\mathbf{m}$ be the intersection of its maximal ideals, and $\{\mathfrak{a}_n\}_{n=1}^{\infty}$ be a decreasing sequence of ideals such that $\cap_{n=1}^{\infty}\mathfrak{a}_n = 0$. Then there exists an integral valued function $s(n)$ which tends to infinity with n, such that $\mathfrak{a}_n \subseteq \mathbf{m}^{s(n)}$.*

Proof. See [ZS60, Chap 8, §5, Thm 13]. □

Lemma 6.26. *Let $\{\mathfrak{a}_n\}_{n=1}^{\infty}$ be a decreasing sequence of ideals in I such that $\cap_{n=1}^{\infty}\mathfrak{a}_n = 0$. Then the natural map $\eta : I \to \varprojlim_n I/\mathfrak{a}_n$ is an isomorphism.*

Proof. It is clear that $\ker(\eta) = \cap_{n=1}^{\infty} \mathfrak{a}_n$, and this is zero by assumption. For the surjectivity, we use Lemma 6.25. Let $(\bar{a}_n)$ be an element of $\varprojlim_n I/\mathfrak{a}_n$. We need to prove that there exists an element $a \in I$, such that $\eta(a) = (\bar{a}_n)$, i.e., $a - a_n \in \mathfrak{a}_n$, $\forall\, n \in \mathbb{N}$. For every $n \in \mathbb{N}$ and $j \geq n$, we have

$$a_j - a_n \in \mathfrak{a}_n.$$

By the above lemma, $a_j - a_n \in \mathbf{m}^{s(n)}$, i.e., the sequence $(\bar{a}_n)$ is a Cauchy sequence with respect to the $\mathbf{m}$-adic topology on I. Since I is complete with respect to this topology, the sequence $\{a_n\}$ converges to an element, say a, i.e., for every $j \geq 1$, we have

$$a - a_j \in \mathbf{m}^{s(j)}.$$

Now we show that, for every $n \in \mathbb{N}$, $a - a_n \in \mathfrak{a}_n$. Let us fix n. Adding the two displays above for $j \geq n$, we see $a - a_n \in \cap_{j \geq n}(\mathfrak{a}_n + \mathbf{m}^{s(j)})$. The ideal $\mathfrak{a}_n$ is closed in the $\mathbf{m}$-adic topology of I [ZS60, Chap 8, §4, Thm. 9] and hence $\mathfrak{a}_n = \cap_{j \geq n}(\mathfrak{a}_n + \mathbf{m}^{s(j)})$. Therefore, $a - a_n \in \mathfrak{a}_n$, for every $n \in \mathbb{N}$. □

Lemma 6.27. *Let R be a commutative noetherian integral domain. Let S denote a countable collection of height 1 prime ideals in R. Then*

$$\bigcap_{P \in S} P = 0.$$

Proof. Denote $\cap_{P \in S} P$ by J. In order to show $J = 0$, it suffices to show that any non-zero element x of J is contained in only finitely many height 1 prime ideals of R. Since R is noetherian, primary decomposition exists for ideals, and we can write

$$\mathrm{Rad}(x) = \cap_{i=1}^{n} Q_i,$$

where $\mathrm{Rad}(x)$ is the radical of (x) and $\{Q_i\}_{i=1}^{n}$ are prime ideals of R. Observe that the Q_i's are non-zero, because x is non-zero. Then

$$\{P \mid P \text{ is a prime ideal of height 1 containing } x\} \subseteq \{Q_1, Q_2, \ldots, Q_n\}.$$ □

We now have:

Theorem 6.28. *Suppose $\{P_n\}_{n=1}^{\infty}$ is a sequence of height 1 prime ideals of I. For each $n \geq 1$, suppose $\pi(P_n)$ is a pseudo-representation of G into I/P_n. Suppose that the $\pi(P_n)$ are compatible on a dense subset Σ of G, i.e., there exist two functions T and D on Σ, with values in I, such that for any $n \geq 1$,*

$$\mathrm{Tr}(\pi(P_n))(\sigma) \equiv T(\sigma) \pmod{P_n},$$
$$\mathrm{Det}(\pi(P_n))(\sigma) \equiv D(\sigma) \pmod{P_n}, \tag{6.29}$$

for all $\sigma \in \Sigma$. Let $\mathfrak{a}_n = \cap_{i=1}^{n} P_i$. Then there exists a pseudo-representation π of G into I, such that for all $\sigma \in \Sigma$, and $n \geq 1$,

$$\mathrm{Tr}(\pi)(\sigma) \equiv T(\sigma) \pmod{\mathfrak{a}_n},$$
$$\mathrm{Det}(\pi)(\sigma) \equiv D(\sigma) \pmod{\mathfrak{a}_n}, \tag{6.30}$$

and hence for every $\sigma \in \Sigma$,

$$\mathrm{Tr}(\pi)(\sigma) = T(\sigma), \ \mathrm{Det}(\pi)(\sigma) = D(\sigma).$$

Proof. First, we shall show that, for every $n \geq 1$, there are pseudo-representations $\pi(\mathfrak{a}_n) : G \to I/\mathfrak{a}_n$ such that

$$\mathrm{Tr}(\pi(\mathfrak{a}_n))(\sigma) \equiv T(\sigma) \pmod{\mathfrak{a}_n}, \tag{6.31}$$

for every $\sigma \in \Sigma$. We do this by induction on n. For $n = 2$, applying Lemma 6.23 for the ideals P_1, P_2, and their associated pseudo-representations $\pi(P_1)$, $\pi(P_2)$, we get that $\pi(\mathfrak{a}_2)$ exists with the property (6.31). Assume that, we can construct pseudo-representations till $n - 1$, i.e., that we can construct $\pi(\mathfrak{a}_1), \ldots, \pi(\mathfrak{a}_{n-1})$ satisfying (6.31). Now, applying Lemma 6.23 for the ideals $\mathfrak{a}_{n-1}$, P_n and their associated pseudo-representations $\pi(\mathfrak{a}_{n-1})$, $\pi(P_n)$, we get that $\pi(\mathfrak{a}_n)$ exists satisfying (6.31). Hence, we have constructed $\pi(\mathfrak{a}_n)$ satisfying (6.31), for all n.

In order to define the pseudo-representation π, let us first define its trace $\mathrm{Tr}(\pi)$ as follows:

$$\mathrm{Tr}(\pi) : G \to \prod_{n=1}^{\infty} I/\mathfrak{a}_n$$

$$g \mapsto (\mathrm{Tr}(\pi(\mathfrak{a}_n))(g)). \tag{6.32}$$

By (6.31), all the $\pi(\mathfrak{a}_n)$'s are compatible (in the sense defined in Lemma 6.23), and hence $\mathrm{Tr}(\pi)(\Sigma) \subseteq \varprojlim_n I/\mathfrak{a}_n$. Now, by Lemma 6.26, $\mathrm{Tr}(\pi)(\Sigma) \subseteq I$, and satisfies:

$$\mathrm{Tr}(\pi)(\sigma) \equiv \mathrm{Tr}(\pi(\mathfrak{a}_n))(\sigma) \pmod{\mathfrak{a}_n}, \tag{6.33}$$

for all $\sigma \in \Sigma$. Since $I = \varprojlim_n I/\mathfrak{a}_n$ is closed in $\prod_{n=1}^{\infty} I/\mathfrak{a}_n$, we get $\mathrm{Tr}(\pi)(G) \subseteq I$, by the continuity of trace and the density of Σ in G. Now, we can construct the map π, from $\mathrm{Tr}(\pi)$, by defining $A(g)$, $D(g)$ as in (6.20). It is easy to check that π is a pseudo-representation by using the fact that each $\pi(\mathfrak{a}_n)$ is a pseudo-representation. By (6.31) and (6.33), it is easy to see that (6.30) holds. The last statement follows from Lemma 6.27. A similar argument works for the Det function. □

Corollary 6.34. *Under the same assumptions as above, the pseudo-representation $\pi : G \to I$, thought of as taking values in K, can be lifted to a representation ρ, such that the trace of ρ is equal to the trace of π.*

Proof. This follows from Corollary 6.22 and Remark 6.19. □

6.7 *Existence of the representation ρ_F*

In this section, given F as in §6.1, we shall show the existence of the representation $\rho = \rho_F$. The strategy is to specialize F at a countable subset of prime ideals P in $\Xi(I)$ for which one has an associated Galois representation by Corollary 6.13, and then to patch these representations to construct ρ.

Theorem 6.35 (Wiles). *Suppose that $\{P_n\}_{n=1}^{\infty}$ is an infinite set of distinct prime ideals of I of height one. Let $\widetilde{I/P_n}$ denote the integral closure of I/P_n in $Q(I/P_n)$. Suppose that, for each n, we are given a continuous, odd representation*

$$\rho_n : \mathrm{G}_{\mathbb{Q}} \to \mathrm{GL}_2(\widetilde{I/P_n}),$$

which is unramified outside Np. Suppose that for each prime $q \nmid Np$, there exist elements a_q and ϵ_q, in I, such that,

$$\mathrm{Trace}\ \rho_n(\mathrm{Frob}_q) = a_q \pmod{P_n}, \text{ and}$$
$$\mathrm{Det}\ \rho_n(\mathrm{Frob}_q) = \epsilon_q \pmod{P_n}.$$

Then there exists a continuous, odd, representation

$$\rho : \mathrm{G}_{\mathbb{Q}} \to \mathrm{GL}_2(K),$$

with

$$\mathrm{Trace}\ \rho(\mathrm{Frob}_q) = a_q,$$
$$\mathrm{Det}\ \rho(\mathrm{Frob}_q) = \epsilon_q,$$

for $q \nmid Np$. Also, ρ is irreducible and unique, if ρ_n is irreducible for some n.

Proof. Let c denote complex conjugation with determinant -1 in all ρ_n's. Since $\widetilde{I/P_n}$ is a discrete valuation ring, we can pick a basis of $\widetilde{I/P_n}^2$ such that $\rho_n(c) = \begin{pmatrix} 1 & 0 \\ 0 & -1 \end{pmatrix}$. We fix a representation ρ_n with this property for each n. Now let

$$\rho_n(\sigma) = \begin{pmatrix} a_\sigma^{(n)} & b_\sigma^{(n)} \\ c_\sigma^{(n)} & d_\sigma^{(n)}, \end{pmatrix}$$

for each $\sigma \in \mathrm{G}_{\mathbb{Q}}$. Let π_n denote the pseudo-representation associated to ρ_n (see Definition 6.17 in §6.6). A priori π_n is a pseudo-representation from $\mathrm{G}_{\mathbb{Q}}$ to $\widetilde{I/P_n}$. What is important for us is that ρ_n is actually a pseudo-representation into I/P_n. Let us explain this point more clearly.

In general, if $\rho(c) = \begin{pmatrix} 1 & 0 \\ 0 & -1 \end{pmatrix}$, then the pseudo-representation ρ only depends on Trace(ρ). In our case, by assumption, Trace(ρ_n) belongs to I/P_n. Hence, we have a sequence of pseudo-representations from $\mathrm{G}_{\mathbb{Q}}$ into I/P_n. Take $\Sigma := \{\mathrm{Frob}_q | q \nmid Np\}$. We know, by the Chebotarev density theorem, that Σ is dense $\mathrm{G}_{\mathbb{Q}}$. We are now in a position to apply Theorem 6.28 to get a pseudo-representation π from $\mathrm{G}_{\mathbb{Q}}$ into I such that

$$\mathrm{Tr}(\pi)(\mathrm{Frob}_q) = a_q \text{ and } \mathrm{Det}(\pi)(\mathrm{Frob}_q) = \epsilon_q,$$

for all $q \nmid Np$. Now think of π as taking values in K. Then, by Corollary 6.34, we can lift this pseudo-representation π to a representation ρ such that both have the same traces and determinants on Σ. It is clear that ρ is irreducible, if ρ_n is irreducible for some n. Also, the representation ρ^{ss} is unique, by Lemma 6.8. Hence the theorem is proved. □

Remark 6.36. The above theorem can be found in a more general form in [Wil88, Lem. 2.2.3], where one works outside an analytic density zero set of primes.

As a consequence of the above theorem, we can finally prove Theorem 6.1:

Proof. We can take a countable collection of prime ideals of I, say $\{P_n\}_{n=1}^{\infty}$, of height 1 from $\Xi(I)$. We can take ρ_n to be the representation associated to $F(P_n)$, which exists by Corollary 6.13. Maintaining the same notations as above, for every prime $q \nmid Np$, take

$$a_q = \lambda(T_q) \in I \quad \text{and} \quad \epsilon_q = (\chi\kappa\nu_p^{-1})(q) \in I.$$

By the computations in (6.14), for every $q \nmid Np$, a_q and ϵ_q satisfy the conditions mentioned in the above theorem. Hence, we have a representation ρ, such that for each prime $q \nmid Np$

$$\det(1 - \rho(\mathrm{Frob}_q)T) = 1 - \lambda(T_q)T + (\chi\kappa\nu_p^{-1})(q)\, T^2,$$

We call this ρ as ρ_F. Hence, we have proved Theorem 6.1. □

Remark 6.37. The same proof works for the prime $p = 2$, with a slight modification in the definition of the pseudo-representation. For further details, see [Wil88].

This finishes the proof of Theorem 6.1. Let us now prove Theorem 6.2.

6.8 *Proof of Theorem 6.2*

Define a continuous semi-simple representation ρ_0 from D_p to $\mathrm{GL}_2(I)$ as follows: for $g \in D_p$ set

$$\rho_0(g) = \begin{pmatrix} A(g) & 0 \\ 0 & D(g) \end{pmatrix},$$

where $A = (\chi\kappa\nu_p^{-1}) \cdot \lambda(a_p(F))^{-1}$ and $D = \lambda(a_p(F))$. Note that A and D take values in I since $a_p(F)$ is a unit in I. Indeed, $a_p(F)$ is a unit in I/P_i, for infinitely many height 1 prime ideals $\{P_i\}_{i=1}^{\infty}$ of I, and $I^{\times} = \varprojlim_n (I/\cap_{i=1}^{n} P_i)^{\times}$.

We shall show that ρ_0 and $(\rho_{F|D_p})^{\mathrm{ss}}$ have the same traces. To prove this, the following observations are useful:

- Let ρ be a representation from $\mathrm{G}_{\mathbb{Q}}$ to $\mathrm{GL}_2(K)$ and let P be any height 1 prime ideal of I. Take a $\mathrm{G}_{\mathbb{Q}}$-stable lattice $\mathcal{L}$ over I. With respect to $\mathcal{L} \otimes_I I_P$, we may view ρ as a representation $\rho : \mathrm{G}_{\mathbb{Q}} \to \mathrm{GL}_2(I_P)$. Let $\rho_P = \rho$ modulo PI_P. (The residual representation $\rho(P)$ of ρ at P, is by definition, the semi-simplification of ρ_P.) Similarly, define $\rho_{|D_p,P}$ to be $\rho_{|D_p}$ modulo PI_P. Clearly

$$\rho_{|D_p,P} = \rho_{P|D_p}. \tag{6.38}$$

- It is known that Deligne's representation $\rho_{F(P)}$ is equivalent to the residual representation $\rho(P)$ attached to $\rho = \rho_F$ at an arithmetic prime P, i.e., $\rho_{F(P)} \sim \rho(P)$. Since Deligne's representation $\rho_{F(P)}$ is irreducible, we further have $\rho_{F(P)} \sim \rho_P$. In particular, we have

$$\rho_{F(P)|D_p} \sim \rho_{P|D_p}. \tag{6.39}$$

Returning to the proof, let $\{P_n\}_{n=1}^{\infty}$ be a sequence of arithmetic height 1 prime ideals of I. Since the traces of ρ_0 and $(\rho_{F|D_p})^{ss}$ belongs to I, it is enough show that they are equal mod P_n, because $\cap_{n\geq 1} P_n = 0$, by Lemma 6.27. But this follows, since mod P_n, we have:

$$\mathrm{trace}(\rho_0) \overset{(6.14)}{\equiv} \mathrm{trace}(\rho_{F(P_n)|D_p}) \overset{(6.39)}{=} \mathrm{trace}(\rho_{P_n|D_p}) \overset{(6.38)}{=} \mathrm{trace}(\rho_{|D_p,P_n}).$$

Thinking of ρ_0 as a representation into $\mathrm{GL}_2(K)$, we conclude that $\rho_0 \sim (\rho_{F|D_p})^{ss}$, by Lemma 6.8.

By using the above description of the representation $(\rho_{F|D_p})^{ss}$, there are two possibilities for the representation $\rho_{F|D_p}$, up to isomorphism, namely:

(1) $\rho_{F|D_p} \sim \begin{pmatrix} A & U \\ 0 & D \end{pmatrix}$, or,

(2) $\rho_{F|D_p} \sim \begin{pmatrix} A & 0 \\ U & D \end{pmatrix}$,

where U is a continuous map. These cases are not mutually exclusive. If the first case holds then we immediately obtain Theorem 6.2. In the second case, we shall show that $\rho_{F|D_p}$ is semi-simple (i.e., U can be taken to be 0), proving the theorem in this case as well.

Indeed, suppose that $\rho_{F|D_p}$ is non-split (i.e., not semi-simple). In particular, one knows that F is not of CM type [GV04, Prop. 12]. By an argument on [GV04, §4, p. 2161], we know that $\rho_{F(P_n)|D_p}$ does not split for all but finitely many specializations. We now show that all the $\rho_{F(P_n)|D_p}$ split, an obvious contradiction. Hence, $\rho_{F|D_p}$ must be split, as desired.

Lemma 6.40. *For an arithmetic prime P of I, if the representation $\rho_{F|D_p}$ is an extension of a ramified character by an unramified character, so is $\rho_{F(P)|D_p}$.*

Proof. By (6.39), it is enough to show that $\rho_{P|D_p}$ has an unramified subspace. This can be shown as follows.

By assumption, there exists a basis, say v_1, v_2, of K^2 such that $\rho_{F|D_p}$ acts K-linearly on v_2 by the character D. Take any $G_{\mathbb{Q}}$-stable lattice $\mathcal{L}$ over I_P. Since I_P is a DVR, there exists an I_P basis w_1, w_2 of $\mathcal{L}$, i.e., $\mathcal{L} = I_P w_1 + I_P w_2$. Now look at $Kv_2 \cap \mathcal{L}$. There exists an integer $r \in \mathbb{Z}$, such that $\pi^r v_2 = aw_1 + bw_2$, where π is a uniformizer of I_P, and $a, b \in I_P$ with $aI_P + bI_P = 1$. Clearly $\pi^r v_2 \in Kv_2 \cap \mathcal{L}$. It is clear that $I_P(\pi^r v_2) \subseteq I_P w_1 + I_P w_2$. Since $aI_P + bI_P = 1$, we can find a vector $\tilde{v}_1 \in I_P w_1 + I_P w_2$, such that $I_P(\pi^r v_2) + I_P \tilde{v}_1 = I_P w_1 + I_P w_2 = \mathcal{L}$.

Let us look at the representation ρ_F with respect to the new lattice $\mathcal{L}_2 = I_P \tilde{v}_1 + I_P(\pi^r v_2)$. It is clear that $\mathcal{L}_2$ is $G_{\mathbb{Q}}$-stable lattice over I_P. With respect to $\mathcal{L}_2$, the representation $\rho_{F|D_p}$ looks like

$$\begin{pmatrix} A_1 & 0 \\ U_1 & D \end{pmatrix} \in \mathrm{GL}_2(I_P). \tag{6.41}$$

Since trace of $\rho_{F|_{D_p}}$ is $A+D$, it is easy to see that $A_1 = A$. Now we have the following isomorphisms/equalities over $Q(I/P)$,

$$\rho_{F(P)|_{D_p}} \overset{(6.39)}{\sim} \rho_{P|_{D_p}} \overset{(6.38)}{=} \rho_{|_{D_p},P} \overset{(6.41)}{=} \begin{pmatrix} \bar{A} & 0 \\ \bar{U}_1 & \bar{D} \end{pmatrix},$$

where bar denotes mod P reduction. Since D is unramified, $\bar{D}$ is also unramified. Also since P is an arithmetic prime, $\bar{A}\bar{D}$ is ramified by (6.12), noting $k \geq 2$ there. Hence, $\bar{A}$ is ramified. This shows that the representation $\rho_{F(P)|_{D_p}}$ is an extension of a ramified character by an unramified character. □

By the above lemma, $\rho_{F(P_n)|_{D_p}}$ is an extension of a ramified character by an unramified character. But by (6.12), it is also an extension of an unramified character by a ramified character. Hence the $\rho_{F(P_n)|_{D_p}}$ split, for all n. This completes the proof of Theorem 6.2.

7. Constructing cusp forms out of semi-cusp forms

Definition 7.1. *A classical form f (respectively Λ-adic form F) is said to be a semi-cusp form if the constant term $a_0(f) = 0$ (respectively $a_0(F) = 0$).*

When there is more than one cusp, a semi-cusp form, whether classical or Λ-adic, is not necessarily a cusp form. At one stage in the proof of the Iwasawa main conjecture (see [Wil90, sec. 3]), Wiles needs to create a Λ-adic cusp form out of a Λ-adic semi-cusp form F. We describe a criterion in [Wil90, sec. 3] which achieves this. This criterion is used at a key point in the next section.

Let ϕ denote the Euler ϕ-function. Define the following Hecke operator (cf. [Wil90, p. 506]):

$$w := e \cdot T_N \cdot \prod_{\ell | N} (T_\ell^{\phi(Np)} - (\kappa(\langle \ell \rangle) l^{-1})^{\phi(Np)}).$$

Remark 7.2. In Wiles' definition of w on [Wil90, p. 506], there is a Λ-adic diamond operator $\langle \ell \rangle$ which is essentially our $\kappa(\langle \ell \rangle)\ell$. The discrepancy of ℓ^2 in our definition of w comes from the fact that in this section we are working with the usual normalization $X \mapsto \zeta u^k - 1$ and not with Wiles' normalization $X = \zeta u^{k-2} - 1$. However, in the next section, we use Wiles' normalization, and so need to use the operators w with $\kappa(\langle \ell \rangle)\ell$, rather than $\kappa(\langle \ell \rangle)\ell^{-1}$. Also, strictly speaking, in our notation we should really be writing U_ℓ instead of T_ℓ.

In any case, we have the following result [Wil90, Lem. 3.2]:

Proposition 7.3. *Suppose p is odd and χ is a primitive character of level Np. If $F \in M(N, \chi, I)$ is a semi-cusp form, i.e., $a_0(F) = 0$, then wF is a cusp form.*

Proof. The proof we give is more explicit that the one given in [Wil90]. We take advantage of the assumption made in these notes that the base field is $\mathbb{Q}$.

Let f_ν be the specialization of F at ν, lying over $\nu_{k,\zeta}$. We must show that wf_ν is a cusp form. Recall that $f_\nu \in M_k(Np^r, \chi_\nu)$, and this last space has a Hecke-stable decomposition:

$$M_k(Np^r, \chi_\nu) = S_k(Np^r, \chi_\nu) \oplus \mathcal{E}(Np^r, \chi_\nu),$$

where $\mathcal{E}(Np^r, \chi_\nu)$ is the Eisenstein part of $M_k(Np^r, \chi_\nu)$. Thus, we may accordingly write:

$$f_\nu = g_\nu + E_\nu.$$

Since f_ν is a semi-cusp form by hypothesis, we note E_ν has the important property that the constant term of E_ν vanishes. Applying w to the above decomposition, we get

$$wf_\nu = wg_\nu + wE_\nu.$$

Since wg_ν is a cusp form, we are reduced to showing that $wE_\nu = 0$. We do this by analyzing the various possibilities for the basis elements that occur when one expands E_ν in a given basis, and by showing that (a part of) w kills each such basis element.

Recall that $\mathcal{E}_k(Np^r, \chi_\nu)$, for $k \geq 2$, has the well-known basis (see, for instance, Section 4.7 of Miyake's book *Modular Forms*):

$$E_k(\chi_1, \chi_2)(mz)$$

where

- χ_1 and χ_2 are characters with $\chi_1\chi_2 = \chi_\nu$, and,
- If $M_1 = \mathrm{cond}(\chi_1)$ and $M_2 = \mathrm{cond}(\chi_2)$, then $mM_1M_2 \mid Np^r$.

When $k = 2$ and both χ_1 and χ_2 are trivial, the basis element $E_2(\chi_1, \chi_2)$ has to be modified slightly: in this case instead of $M_1 = M_2 = 1$, one takes $M_1 = 1$ and M_2 to be a prime number. Here $E_k(\chi_1, \chi_2)$ is the Eisenstein series defined by the identities:

$$L(E_k(\chi_1, \chi_2), s) = L(s, \chi_1)L(s - k + 1, \chi_2),$$

that is,

$$a_n(E_k(\chi_1, \chi_2)) = \sum_{d|n} \chi_1\left(\frac{n}{d}\right)\chi_2(d)d^{k-1}, \text{ for } n \geq 1,$$

and,

$$a_0(E_k(\chi_1, \chi_2)) = \begin{cases} 0 & \text{if } \chi_1 \neq 1, \\ \frac{-B_{k,\chi}}{2k} & \text{otherwise (except it's} \\ \frac{M_2-1}{24} & \text{when } k = 2, \text{ and } \chi_1, \chi_2 \text{ are trivial).} \end{cases}$$

In particular $E_k(\chi_1, \chi_2)(mz)$ has vanishing constant term iff $\chi_1 \neq 1$.

Write E_ν as a linear combination of the above basis elements. Since E_ν has zero constant term, not all the basis elements can occur. If the p-part of $\chi_\nu = \chi\omega^{-k}\chi_\zeta$ is primitive (e.g., if $r > 1$ or the p-part of χ is not ω^k), then χ_ν is primitive of level Np^r, so $m = 1$, and there is only one basis element with non-zero constant term, namely $E_k(1, \chi_\nu)(z)$. Thus, the only basis elements that can occur in E_ν are the $E_k(\chi_1, \chi_2)(z)$ with $\chi_1 \neq 1$. Now assume that $r = 1$ and the p-part of χ_ν is trivial. Think of χ_ν as a character of level N, and note $E_k(1, \chi_\nu)$ has level N (instead of Np). In this case there are two basis elements with non-zero constant term, namely

$E_k(1, \chi_\nu)(z)$ and $E_k(1, \chi_\nu)(pz)$. Change basis and replace these forms by their p-stabilizations:

$$E_k^{(p)}(1, \chi_\nu)(z) = E_k(1, \chi_\nu)(z) - \chi_\nu(p)p^{k-1}E_k(1, \chi_\nu)(pz)$$
$$F_k^{(p)}(1, \chi_\nu)(z) = E_k(1, \chi_\nu)(z) - E_k(1, \chi_\nu)(pz).$$

These have p-th Fourier coefficients 1, and $\chi_\nu(p)p^{k-1}$, respectively. Clearly $E_k^{(p)}$ has non-zero constant term, whereas $F_k^{(p)}$ has zero constant term. Thus again in this new basis the only basis elements that occur in this linear combination of E_ν are the $E_k(\chi_1, \chi_2)(mz)$ with $\chi_1 \neq 1$ and $m = 1$ or p, and $F_k^{(p)}$.

Thus it suffices to show that w kills the basis elements $E_k(\chi_1, \chi_2)(mz)$ with $\chi_1 \neq 1$, and $F_k^{(p)}$. We do this case by case. First note that Hida's idempotent e, hence w, kills $F_k^{(p)}$, since $a_p(F_k^{(p)})$ is not a p-adic unit. Similarly if $m > 1$, then $a_p(E_k(\chi_1, \chi_2)(mz)) = 0$, so again w kills $E_k(\chi_1, \chi_2)(mz)$. Thus it suffices to show that w kills the basis elements $E_k(\chi_1, \chi_2)(z)$ with $\chi_1 \neq 1$, i.e., with $M_1 \neq 1$. We have the further cases:

(1) Suppose $\ell | M_1$ and $\ell | M_2$, for some $\ell | Np^r$: In this case $a_\ell = \chi_1(\ell) + \chi_2(\ell)\ell^{k-1} = 0$. If $\ell | N$, then $a_\ell | a_N$ and $a_N = 0$, so T_N kills $E_k(\chi_1, \chi_2)$. If $\ell = p$, then $a_p = 0$ in which case e does the trick.

(2) Suppose $\ell | M_1$ but $\ell \nmid M_2$, for some $\ell | Np^r$. Write E_k for $E_k(\chi_1, \chi_2)$, for simplicity. There are two further sub-cases:

 (i) Suppose $\ell \neq p$. In this case $a_\ell = \chi_1(\ell) + \chi_2(\ell)\ell^{k-1} = \chi_2(\ell)\ell^{k-1}$. Thus

$$T_\ell E_k = a_\ell E_k = \epsilon_1 l^{k-1} E_k,$$

 where ϵ_1 is a $\phi(N)$-th root of unity. Now as we have checked many times before $\kappa(\langle \ell \rangle)l^{-1}$ specializes under ν to $\omega^{-k}\chi_\zeta(\ell)l^{k-1}$, so

$$(\kappa(\langle \ell \rangle)l^{-1})E_k = \epsilon_2 \zeta^{s(\langle \ell \rangle)} l^{k-1} E_k,$$

 for some $\phi(p)$-th root of unity ϵ_2. Thus:

$$(T_\ell^{\phi(Np)} - (\kappa(\langle \ell \rangle)l^{-1})^{\phi(Np)})E_k = (1 - \zeta')(l^{k-1})^{\phi(Np)} E_k,$$

 for some p-power root of unity ζ'. But note that $1 - \zeta'$ is not a p-adic unit (though the power of ℓ is), so further composing with e kills E_k. (NB: The last part of the argument just given corrects the somewhat incorrect hint on [Wil90, p. 506, line 19], which claims that the previous line is already identically 0; this seems to us to hold only in the special case when ζ and hence ζ' are 1).

 (ii) Finally suppose $\ell = p$. Then $a_\ell = a_p = \chi_2(p)p^{k-1}$ is not a p-adic unit, so again $eE_k = 0$.

Since, in all cases w kills the basis elements in E_ν, we are done. □

8. Eisenstein ideals and p-adic L-functions

In Mazur's famous paper [Maz77], key use is made of the so called Eisenstein ideal. This ideal measures congruences between cusp forms of weight 2 for $\Gamma_0(p)$ and the (unique, normalized) Eisenstein series of weight 2 for $\Gamma_0(p)$. As with many things in [Wil90], Wiles defines a Λ-adic analogue of this ideal (cf. [Wil90, Sec. 4]), and proves a result which shows that it measure congruences between Λ-adic cusp forms (of some tame level N and character χ), and the Λ-adic Eisenstein series $\mathcal{E}_\chi$ introduced in Section 4. We describe Wiles' result in this section.

8.1 *Classical Eisenstein ideal*

As motivation, we first recall some facts about the classical Eisenstein ideal.

We have the standard decomposition of forms of level p and trivial nebentypus:

$$M_2(p, 1) = S_2(p, 1) \oplus \mathcal{E}_2(p, 1),$$

where the space of cusp forms $S_2(p, 1)$ has dimension equal to the genus of $X_0(p)$, and $\mathcal{E}_2(p, 1)$ is one-dimensional spanned by the Eisenstein series that was earlier called $E_2^{(p)}$ (this is also $E_2(\chi_1, \chi_2)$ of the last section, with χ_1 trivial of conductor 1, and χ_2 trivial of conductor p). As mentioned in the last section, $E_2^{(p)}$ has q-expansion:

$$E_2^{(p)} = \frac{p-1}{24} + \sum_{n=1}^{\infty} \sigma_1^{(p)}(n)\, q^n.$$

One is interested in studying congruences between cuspidal eigenforms f_2 and $E_2^{(p)}$, i.e., congruences of the form

$$f_2 \equiv E_2^{(p)} \bmod \ell,$$

where ℓ is a prime (call such primes ℓ Eisenstein congruence primes). Heuristically, one can write down such congruences by noting that if ℓ divides the constant term of $E_2^{(p)}$, then $E_2^{(p)}$ mod ℓ looks like a mod ℓ cuspidal eigenform, and so should lift to a characteristic zero cuspidal eigenform. This is in fact what happens. To make this precise, we need the notion of a congruence module. Let $\mathbb{T}_{M_2}$, respectively, $\mathbb{T}_{S_2}$, $\mathbb{T}_{\mathcal{E}_2}$, be the Hecke algebras acting on the three spaces in the decomposition above. There are natural projection maps $\mathbb{T}_{M_2} \twoheadrightarrow \mathbb{T}_{S_2}$ and $\mathbb{T}_{M_2} \twoheadrightarrow \mathbb{T}_{\mathcal{E}_2}$, and these induce the natural map $\mathbb{T}_{M_2} \hookrightarrow \mathbb{T}_{S_2} \oplus \mathbb{T}_{\mathcal{E}_2}$.

Definition 8.1. *Let*

$$C(\mathbb{T}) = \frac{\mathbb{T}_{S_2} \oplus \mathbb{T}_{\mathcal{E}_2}}{\mathbb{T}_{M_2}}$$

be the congruence module with respect to the decomposition above.

It is easy to see that $|C(\mathbb{T})| < \infty$ and $\ell \mid |C(\mathbb{T})|$ iff there is a congruence of the form we are interested in. Thus the Eisenstein congruence primes are exactly the primes in the support of $|C(\mathbb{T})|$.

There is another way to write down $C(\mathbb{T})$, that is completely intrinsic to $\mathbb{T}_{S_2}$. This uses the Eisenstein ideal I, an ideal of $\mathbb{T}_{S_2}$. Noting that $a_q(E_2^{(p)}) = q + 1$ for primes $q \neq p$ and $a_p(E_2^{(p)}) = 1$, one defines:

Definition 8.2. *The Eisenstein ideal I is the ideal of T_{S_2} spanned by $T_q - (q+1)$ for all $q \neq p$, and by $U_p - 1$.*

Lemma 8.3. *The natural map $\mathbb{T}_{S_2} \hookrightarrow \mathbb{T}_{S_2} \oplus \mathbb{T}_{\mathcal{E}_2}$ induces an isomorphism*

$$\frac{\mathbb{T}_{S_2}}{I} \xrightarrow{\sim} C(\mathbb{T}).$$

In view of the lemma we deduce that the Eisenstein congruence primes are those in the support of T_{S_2}/I. But now, we have (cf. [Maz77, Prop. 9.7]):

Theorem 8.4. *Let n be the numerator of $\left(\frac{p-1}{12}\right)$. Then*

$$\frac{\mathbb{T}_{S_2}}{I} \xrightarrow{\sim} \frac{\mathbb{Z}}{n}.$$

In particular we see that $n \in I$ (compare with the definition of I below).

One concludes that the Eisenstein congruence primes are exactly the prime divisors of the numerator of (twice) the constant term of $E_2^{(p)}$.

8.2 Λ-*adic Eisenstein ideal*

Following Wiles, we wish to 'Λ-fy' the previous discussion. Let p be an odd prime and let χ be a fixed non-trivial, even, primitive Dirichlet character of level Np. In Section 4, we had introduced the Λ-adic Eisenstein series $\mathcal{E}_\chi$ attached to χ. This has q-expansion in $I_\chi[[q]]$ where $I_\chi = \mathbb{Z}_p[\chi][[X]]$ (not to be confused with I, which in this section is reserved for the Eisenstein ideal). To avoid errors in what follows we work with Wiles' normalization, so in particular, we take $\mathcal{E}_\chi$ as modified in Section 4.5. Consider the direct sum

$$S^{\mathrm{ord}}(N, \chi, I_\chi) \oplus I_\chi \cdot \mathcal{E}_\chi$$

inside the space $M^{\mathrm{ord}}(N, \chi, I_\chi)$. As before, we wish to understand congruences between ordinary Λ-adic cuspidal eigenforms, and $\mathcal{E}_\chi$. Let $\mathbb{T}_{S^{\mathrm{ord}}}$ denote the Λ-adic Hecke algebra acting on $S^{\mathrm{ord}}(N, \chi, I_\chi)$. Noting that $a_n(\mathcal{E}_\chi) = A_{n,\chi}(X) \in I_\chi$, for all integers $n \geq 1$ (where $A_{n,\chi}(X)$ is modified as in Section 4.5), we define (cf. [Wil90, eq. (4.2)]):

Definition 8.5. *The Λ-adic Eisenstein ideal is the ideal in $\mathbb{T}_{S^{\mathrm{ord}}}$ generated by*

- *$T_q - A_{q,\chi}(X)$ for prime $q \nmid Np$,*
- *$U_q - A_{q,\chi}(X)$, for prime $q | Np$, and,*
- *$\hat{G}_\chi^0(X)$.*

Recall that $\hat{G}_\chi(X)$ is the numerator of (twice) the constant term of Wiles' $\mathcal{E}_\chi$. Here $\hat{G}^0_\chi(X)$ is the power series constructed from $\hat{G}_\chi(X)$ by dividing out any possible factors of the form $(1 + X - \zeta u^{-1})$ occurring in $\hat{G}_\chi(X)$. As Wiles remarks, one takes the modified p-adic L-function $\hat{G}^0_\chi(X)$ in I in order to avoid certain zeros which cause later complications in his proof of the Iwasawa main conjecture.

The main result of this section is the following theorem (cf. [Wil90, Thm. 4.1]):

Theorem 8.6. *Let χ_0 denote the trivial character of conductor p. Suppose that either $\chi \neq \omega^{-2}$ or $L_p(1, \chi_0) = \infty$. Then*

$$\frac{\mathbb{T}_{S^{\mathrm{ord}}}}{I} \xrightarrow{\sim} \frac{I_\chi}{(\hat{G}^0_\chi(X))}.$$

Before we begin the proof, we need the following abstract lemma.

Lemma 8.7. *Let $\mathfrak{b} = (G^0)$ be a principal ideal of I_χ, and say that $F \in S^{\mathrm{ord}}(N, \chi, I_\chi)$ is a mod $\mathfrak{b}$ eigenform. Assume that $(G^0, a_1(F)) = 1$ so that $a_1(F)$ is invertible in $I_\chi/\mathfrak{b}$. Then the map*

$$\begin{aligned} \lambda : \mathbb{T}_{S^{\mathrm{ord}}} &\to I_\chi/\mathfrak{b} \\ T_m &\mapsto \frac{a_m(F)}{a_1(F)} \end{aligned}$$

is a surjective I_χ-algebra homomorphism.

Proof. When F is normalized, i.e., $a_1(F) = 1$, this is immediate by the correspondence between normalized eigenforms, and homomorphisms of the Hecke algebra. When F is not normalized, we leave the proof that the given map is still a homomorphism as an exercise. □

Proof. Returning to the proof of the theorem, we wish to apply the above lemma with $G^0 = \hat{G}^0_\chi$ and $F = F'$, a certain non-normalized mod $\mathfrak{b}$ cuspidal eigenform of Eisenstein type. We construct F' explicitly now.

Let ψ be an auxiliary even character of conductor Np^r. Let $E_1(1, \psi\omega^{-1})$ be the Eisenstein series defined in the previous section, but for $k = 1$. Write the constant term in lowest terms as $g_{1,\psi\omega^{-1}}/h_{1,\psi\omega^{-1}}$. Set

$$J_{\chi,\psi} := E_1(1, \psi\omega^{-1}) \cdot \mathcal{E}_{\chi\psi^{-1}}(u^{-1}(1 + X) - 1).$$

Here are some remarks.

- $J_{\chi,\psi} \in M(N, \chi, I_\chi)$ is a Λ-adic form in the sense of Wiles. Indeed the first term lies in $M_1(Np^r, \psi\omega^{-1})$ and the second term, after specialization at $X = \zeta u^{k-2} - 1$, lies in $\mathcal{E}_{k-1}(Np^r, \chi\psi^{-1}\omega^{2-(k-1)}\chi_\zeta)$, so the product (after specialization) lies in $M_k(Np^r, \chi\omega^{2-k}\chi_\zeta)$, as desired.
- The constant term of $J_{\chi,\psi}$ is

$$\frac{1}{2} \cdot \frac{g_{1,\psi\omega^{-1}}}{h_{1,\psi\omega^{-1}}} \cdot \frac{\hat{G}_{\chi\psi^{-1}}(u^{-1}(1 + X) - 1)}{\hat{H}_{\chi\psi^{-1}}(u^{-1}(1 + X) - 1)}.$$

Following Wiles, define the following two expressions in I_χ:

$$h_{1,\psi} = \hat{H}_\chi(X) \cdot g_{1,\psi\omega^{-1}} \cdot \hat{G}_{\chi\psi^{-1}}(u^{-1}(1+X)-1),$$
$$h_{2,\psi} = \hat{G}_\chi(X) \cdot h_{1,\psi\omega^{-1}} \cdot \hat{H}_{\chi\psi^{-1}}(u^{-1}(1+X)-1),$$

and set

$$F = 2(h_{1,\psi}\mathcal{E}_\chi - h_{2,\psi} J_{\chi,\psi}).$$

Then by construction F is a Λ-adic semi-cusp form in $M(N, \chi, I_\chi)$, i.e., $a_0(F) = 0$.

By Proposition 7.3, the form $F' = \mathfrak{w}F$ is a cusp form. We use the operator $\mathfrak{w}$ as described in Remark 7.2, since in this section we are using Wiles' normalization. Noting that $h_{2,\psi}$ is divisible by $\hat{G}^0_\chi(X)$, we compute:

$$\begin{aligned}
F' &= 2h_{1,\psi} \cdot \mathfrak{w}\mathcal{E}_\chi - 2h_{2,\psi} \cdot \mathfrak{w}J_{\chi,\psi} \\
&\equiv 2h_{1,\psi} \cdot \left(e \cdot T_N \cdot \prod_{\ell|N} (T_\ell^{\phi(Np)} - (\kappa(\langle \ell \rangle) l)^{\phi(Np)}) \right) \mathcal{E}_\chi \bmod (\hat{G}^0_\chi) \\
&\equiv 2h_{1,\psi} \cdot \left(\prod_{\ell|N} (1 - (\kappa(\langle \ell \rangle) l)^{\phi(Np)}) \right) \mathcal{E}_\chi \bmod (\hat{G}^0_\chi),
\end{aligned}$$

since

$$\begin{aligned}
e\mathcal{E}_\chi &= \mathcal{E}_\chi, \\
T_N\mathcal{E}_\chi &= \mathcal{E}_\chi, \quad \text{and,} \\
T_\ell\mathcal{E}_\chi &= \mathcal{E}_\chi,
\end{aligned}$$

for all $\ell|N$.

Thus if $\mathfrak{b} = (\hat{G}^0_\chi)$, then $F' \equiv c\mathcal{E}_\chi \bmod \mathfrak{b}$, with

$$c = 2h_{1,\psi} \cdot \left(\prod_{\ell|N} (1 - (\kappa(\langle \ell \rangle) l)^{\phi(Np)}) \right).$$

In particular F' is a (non-normalized) mod $\mathfrak{b}$ cuspidal eigenform. To apply Lemma 8.7 we need to further note:

Lemma 8.8. *The following hold:*

1. *There is a ψ such that $h_{1,\psi}$ has no zero in common with $\hat{G}_\chi(X)$, except possibly those of $\hat{H}_\chi(X)$.*
2. *If $\chi \neq \omega^{-2}$ or $L_p(1, \chi_0) = \infty$, then $\hat{G}_\chi(X)$ has no zero in common with $\hat{H}_\chi(X)$.*
3. *For $\ell|N$, the power series $1 - (\kappa(\langle \ell \rangle) l)^{\phi(Np)}$ has no zero in common with $\hat{G}^0_\chi(X)$.*

In particular, for appropriate ψ, which we fix once and for all, and under the hypotheses of the theorem, c has no zero in common with $\hat{G}^0_\chi(X)$.

Proof. We prove the statements one by one.

(1) Let $\psi^{-1} = \chi_\zeta$. Then, a short computation using (4.3) shows that

$$\hat{G}_{\chi\psi^{-1}}(u^{-1}(1+X)-1) = G_\chi(\zeta u^{-1}(1+X)-1).$$

By choosing ζ of sufficiently large order we see that we may assume this power series has no zeros in common with $\hat{G}_\chi(X)$, from which (1) follows.

(2) If $\chi\omega^2 \neq 1$, then $\chi\omega^2$ is not of type W, so $\hat{H}_\chi(X) = H_{\chi\omega^2}(u^2(1+X)-1) = 1$ and the claim follows trivially. If $\chi\omega^2 = 1$ then $\hat{H}_\chi(X) = u^2(1+X)-1$, and the only root is $X = u^{-2}-1$. But on substituting $X = u^{-2}-1$, we have:

$$\frac{\hat{G}_\chi(X)}{\hat{H}_\chi(X)} = \frac{G_{\chi_0}(u^2(1+X)-1)}{H_{\chi_0}(u^2(1+X)-1)} = \frac{G_{\chi_0}(u^0-1)}{H_{\chi_0}(u^0-1)} = L_p(1-0,\chi_0) = \infty,$$

by hypothesis, so the numerator of the LHS does not have a zero at $X = u^{-2}-1$.

(3) Since $\omega(\ell)^{\phi(p)} = 1$ and $\langle \ell \rangle = u^{s(\langle \ell \rangle)}$ we have

$$(\kappa(\langle \ell \rangle)l)^{\phi(Np)} = ((1+X)^{s(\langle \ell \rangle)}\langle l \rangle)^{\phi(Np)} = ((1+X)u)^{s(\langle \ell \rangle)\phi(Np)} = 1$$

iff $X = \zeta u^{-1} - 1$. But these were precisely the zeros that were removed from $\hat{G}_\chi(X)$ to get $\hat{G}^0_\chi(X)$. □

Applying Lemma 8.7 to to $\mathfrak{b} = (\hat{G}^0_\chi)$, and F' with $a_1(F') = c$, which as we have just noted is coprime to $\hat{G}^0_\chi$, we get the last isomorphism of the following sequence of surjective I_χ-algebra homomorphisms:

$$\frac{I_\chi}{(\hat{G}^0_\chi(X))} \twoheadrightarrow \frac{\mathbb{T}_{\mathcal{S}^{\mathrm{ord}}}}{I} \twoheadrightarrow \frac{\mathbb{T}_{\mathcal{S}^{\mathrm{ord}}}}{\ker(\lambda)} \xrightarrow{\sim} \frac{I_\chi}{(\hat{G}^0_\chi(X))}.$$

The second surjection arises by noting that $I \subset \ker(\lambda)$, since by Lemma 8.7, the homomorphism λ takes T_n to $a_n(F')/c \equiv A_{n,\chi} \bmod \mathfrak{b}$. Finally, the first surjection arises from the structure map $I_\chi \to \mathbb{T}_{\mathcal{S}^{\mathrm{ord}}}$ by noting that $\hat{G}^0_\chi(X) \in I$.

But a surjective endomorphism of a noetherian ring is an isomorphism, proving the theorem. □

References

[CS10] J. Coates and R. Sujatha, The main conjecture, This volume.

[Dal10] C. Dalawat, Ribet's modular construction of unramified p-extensions of $\mathbb{Q}(\mu_p)$, This volume.

[Gha05] E. Ghate, Ordinary forms and their local Galois representations, In Algebra and number theory, pages 226–242, Hindustan Book Agency, Delhi (2005).

[GV04] E. Ghate and V. Vatsal, On the local behaviour of ordinary Λ-adic representations, *Ann. Inst. Fourier, Grenoble*, **54(7)** (2004) 2143–2162.

[Hid86] H. Hida, Galois representations into $\mathrm{GL}_2(\mathbf{Z}_p[[X]])$ attached to ordinary cusp forms, *Invent. Math.*, **85** (1986) 545–613.

[Hid93] H. Hida, Elementary Theory of L-functions and Eisenstein series, Cambridge University Press (1993).

[Mat86] H. Matsumura, Commutative ring theory, Cambridge studies in advanced mathematics, **8** (1986).

[Maz77] B. Mazur, Modular curves and the Eisenstein ideal, *Inst. Hautes Études Sci. Publ. Math.*, **47** (1977) 33–186.

[MW86] B. Mazur and A. Wiles, On p-adic analytic families of Galois representations, *Compositio Math.*, **59** (1986) 231–264.

[Rib76] K. Ribet, A modular construction of unramified p-extensions of $\mathbb{Q}(\mu_p)$, *Invent. Math.*, **34(3)** (1976) 151–162.

[Ski04] C. Skinner, CMS lecture notes, Zhejiang University, Hangzhou (2004).

[Ser89] J-P. Serre, Abelian l-adic representations and elliptic curves. Second edition, Advanced Book Classics. Addison-Wesley Publishing Company, Redwood City, CA (1989).

[Ven10] O. Venjakob, Deligne-Ribet's work on L-values, This volume.

[Was96] L. Washington, Introduction to cyclotomic fields, Second edition, Springer-Verlag, Berlin-New York (1996).

[Wil88] A. Wiles, On ordinary λ-adic representations associated to modular forms, *Inv. Math.*, **94(3)** (1988) 529–573.

[Wil90] A. Wiles, The Iwasawa conjecture for totally real fields, *Ann. Math.*, **131** (1990) 493–540.

[ZS60] O. Zariski and A. Samuel, Commutative Algebra, The University Series (1960).

Guwahati Workshop on Iwasawa Theory of Totally Real Fields No. 12, 2010, pp. 49–65.

Ribet's Construction of a Suitable Cusp Eigenform

A. Saikia

Department of Mathematics, IIT Guwahati, Guwahati 781 039, India
e-mail: a.saikia@iitg.ernet.in

1. Preliminaries

We begin by recalling some of the rudiments of modular forms. Other basic ingredients are included in the Appendix.

1.1 *Modular forms*

Let p be an odd prime. Let $\mathfrak{h}$ denote the upper half complex plane, i.e.,

$$\mathfrak{h} = \{z \in \mathbb{C} | Im(z) > 0\}.$$

Let $SL_2(\mathbb{Z})$, $\Gamma_0(p)$ and $\Gamma_1(p)$ respectively denote the following groups:

$$SL_2(\mathbb{Z}) = \left\{ \begin{bmatrix} a & b \\ c & d \end{bmatrix} | a, b, c, d \in \mathbb{Z}, ad - bc = 1 \right\}$$

$$\Gamma_0(p) = \left\{ \begin{bmatrix} a & b \\ c & d \end{bmatrix} \in SL_2(\mathbb{Z}) | c \equiv 0 \text{ modulo } p \right\},$$

$$\Gamma_1(p) = \left\{ \begin{bmatrix} a & b \\ c & d \end{bmatrix} \in \Gamma_0(p) | a \equiv 1 \text{ modulo } p, d \equiv 1 \text{ modulo } p \right\},$$

Let $GL_2(\mathbb{Q})$ ($GL_2(\mathbb{R})$) denote the 2×2 invertible matrices with rational (real) coefficients. It is easy to note all these matrix groups act on $\mathfrak{h}$ by sending z to $\frac{az+b}{cz+d}$. For a function $f : \mathfrak{h} \longrightarrow \mathbb{C}$ and any fixed integer $k \geq 0$, we can define a function $f|[\gamma]_k$ as

$$f|[\gamma]_k(z) = (cz + d)^{-k} f(\gamma(z)) \quad \forall \ \gamma = \begin{bmatrix} a & b \\ c & d \end{bmatrix} \in GL_2(\mathbb{Q}).$$

A function $f : \mathfrak{h} \longrightarrow \mathbb{C}$ is called *weakly modular* of weight k with respect to Γ if $f|[\gamma]_k = f$ for all $\gamma \in \Gamma$ where Γ can mean anyone of $SL_2(\mathbb{Z})$, $\Gamma_0(p)$ or $\Gamma_1(p)$. It is clear that $\begin{bmatrix} 1 & 1 \\ 0 & 1 \end{bmatrix} \in \Gamma$ and hence we must have $f(z + 1) = f(z)$ for a weakly modular function. If f is holomorphic on $\mathfrak{h}$, we can look at the Fourier expansion of f in terms of $q = e^{2\pi i z}$, i.e., $\sum_{n=-\infty}^{+\infty} a_n q^n$. We say f is holomorphic at ∞ if its q-expansion does not involve negative powers of q, i.e., $a_n = 0$ for $n < 0$. If $a_n = 0$ for $n \leq 0$, then we say that f vanishes at ∞. Note that $q = e^{2\pi i z} \to 0$ as $Im(z) \to \infty$, justifying the terminology. We say that f is a modular form of weight k with respect to Γ if

(i) f is weakly modular of weight k with respect to Γ.
(ii) f is holomorphic on $\mathfrak{h}$.
(iii) $f|[\gamma]_k$ is holomorphic at ∞ for all $\gamma \in SL_2(\mathbb{Z})$.
(iv) If, in addition, the q-expansion of $f|[\gamma]_k$ has $a(0) = 0$ for all $\gamma \in \Gamma$, then f is said to be a cusp form.

Note that it is enough to check the last two conditions for a finite number of coset representatives $\{\alpha_i\}$ of Γ in $SL_2(\mathbb{Z})$. The set $\{\alpha_i(\infty)\}$ is known as the *cusps* of Γ. Let us denote the space of all modular forms (cusp forms) of weight k for Γ by $M_k(\Gamma)$ ($S_k(\Gamma)$ respectively). These turn out to be finite dimensional vector spaces. The quotient vector space of $M_k(\Gamma)$ by $S_k(\Gamma)$ is known as the Eisenstein space, denoted by $\mathcal{E}_k(\Gamma)$. It can be identified as the orthogonal complement of $S_k(\Gamma)$ under Petersson inner product, and hence can be thought of as a subspace of $M_k(\Gamma)$ (see section 6.6 of Appendix).

1.2 *Semi-cusp forms*

Definition 1.1. *A semi-cusp form f is a modular form whose leading Fourier coefficient is* 0*, though $f|[\gamma]_k$ need not have its leading Fourier coefficient* 0 *for all $\gamma \in SL_2(\mathbb{Z})$. In other words, a semi-cusp form vanishes at ∞, but it need not vanish at the other 'cusps'. We shall denote the space of semi-cusp forms of Γ by $S'_k(\Gamma)$.*

Consider the map

$$\beta : \Gamma_0(p) \longrightarrow (\mathbb{Z}/p\mathbb{Z})^\times, \quad \gamma = \begin{bmatrix} a & b \\ c & d \end{bmatrix} \mapsto d \bmod p.$$

(Note that $(d, p) = 1$ for $\gamma \in \Gamma_0(p)$ as $ad - bc = 1$ and $p|c$). Clearly, $\Gamma_1(p)$ is the kernel of β, and the quotient is $(\mathbb{Z}/p\mathbb{Z})^\times$. For a character ϵ of $\left(\frac{\mathbb{Z}}{p\mathbb{Z}}\right)^\times$, we can define a subspace $M_k(\Gamma_1(p), \epsilon)$ of $M_k(\Gamma_1(p))$, which consists of modular forms f such that $f|[\gamma]_k = \epsilon(d) f$ for any $\gamma = \left[\begin{smallmatrix} a & b \\ c & d \end{smallmatrix}\right] \in \Gamma_0(p)$. We can define $S'_k(\Gamma_1(p), \epsilon)$ and $S_k(\Gamma_1(p), \epsilon)$ analogously. Note that any character of $\left(\frac{\mathbb{Z}}{p\mathbb{Z}}\right)^\times$ is of the form ω^i, $i = 0, 1, \ldots, (p-2)$ where ω is the Teichmuller character (see section 6.5 Appendix).

1.3 *Examples of modular forms*

For a non-trivial even character ϵ of $\left(\frac{\mathbb{Z}}{p\mathbb{Z}}\right)^\times$, we have the following Eisenstein series of weight 2 and type ϵ (cf chapter 4 of [Di-S]:

$$G_{2,\epsilon} = \frac{L(-1, \epsilon)}{2} + \sum_{n \geq 1} \sum_{d|n} \epsilon(d) d q^n, \tag{1}$$

$$s_{2,\epsilon} = \sum_{n \geq 1} \sum_{d|n} \epsilon\left(\frac{n}{d}\right) d q^n. \tag{2}$$

These two form a basis for the Eisenstein space $\mathcal{E}_2(\Gamma_1(p), \epsilon)$ (cf theorem 4.6.2 [Di-S]). Note that $s_{2,\epsilon}$ is a semi-cusp form. Moreover, both of these are eigenvectors for all Hecke operators T_l with $(l, p) = 1$ (cf proposition 5.2.3 [Di-S]):

$$T_l s_{2,\epsilon} = (l + \epsilon(l)) s_{2,\epsilon}, \qquad T_l G_{2,\epsilon} = (1 + \epsilon(l) l) G_{2,\epsilon}.$$

(See section 6.7 of the Appendix for Hecke operators.)

If ϵ is an odd character of $\left(\frac{\mathbb{Z}}{p\mathbb{Z}}\right)^{\times}$, we have an Eisenstein series of weight 1 and type ϵ given by (cf section 4.8 in [Di-S])

$$G_{1,\epsilon} = \frac{L(0, \epsilon)}{2} + \sum_{n \geq 1} \sum_{d|n} \epsilon(d) q^n.$$

The above three forms have coefficients defined over $\mathbb{Q}(\mu_{p-1})$, where μ_{p-1} denotes the $(p-1)^{th}$ roots of 1. Let $\wp$ denote any of the unramified primes of $\mathbb{Q}(\mu_{p-1})$ lying above p. Clearly, all the Eisenstein forms given above have $\wp$ integral coefficients (except possibly for the constant terms, but see lemma 3.1 later).

For the trivial character $\epsilon = 1$, we have the following Eisenstein series (cf Theorem 4.6.2 in [Di-S]) in $M_k(\Gamma_0(p)) = M_k(\Gamma_1(p), 1)$:

$$G_k = -\frac{B_k}{2k} + \sum_{n \geq 1} \sum_{d|n} d^{k-1} q^n \quad \text{for } k \geq 4, \tag{3}$$

$$G_2 = E_2(z) - pE_2(pz), \quad \text{where } E_2(z) = -\frac{B_2}{4} + \sum_{n \geq 1} \sum_{d|n} d q^n, \tag{4}$$

2. Key steps in the construction of the unramified p-extension

For Ribet's construction of an unramified extension of $\mathbb{Q}(\mu_p)$, it is enough to have a Galois representation on which the Frobenius elements act in a suitable way. We can use the representation associated with a cusp eigenform (cf chapter 9 of [Di-S]). But we need to show that there indeed exists a cusp eigenform whose eigenvalues have certain congruence properties.

The Eisenstein series $G_{2,\epsilon}$ is a simultaneous eigenform for the Hecke operators T_l where l is a prime other than p, with corresponding eigenvalues $1 + \epsilon(l)l \equiv 1 + l^{k-1}$ modulo $\wp$. Here, $\wp$ denotes a prime of $\mathbb{Q}(\mu_{p-1})$ lying above p. It turns out that we need precisely these congruence properties for the Hecke eigenvalues of a *cusp* form. Ribet's idea is to subtract off the constant term of the Eisenstein series $G_{2,\epsilon}$ in a way that preserves the congruence properties of the coefficients and leaves us with a semi-cusp form f which is an eigenvector modulo $\wp$ for all Hecke operators T_l with $(l, p) = 1$. Then one can invoke a result of Deligne and Serre and obtain a semi-cusp form f' which is also an eigenvector for the T_l's with eigenvalues congruent to those of f modulo $\wp$. The congruence properties of f' then ensure that f' is actually a cusp form. Any cusp form in $S_2(\Gamma_1(p))$ is bound to be a newform. Thus, one can invoke the theory of newforms to conclude that f' is in fact a cusp eigenform, that is, an eigenvector for all Hecke operators including T_n's with $p|n$.

To remove the constant term of the Eisenstein series $G_{2,\epsilon}$ without affecting the congruence properties of its coefficients modulo $\wp$, it suffices to produce another Eisenstein series whose constant term is a $\wp$-unit. This will be done in the next section.

3. Construction of an Eisenstein series with $\wp$-unit constant term

As before, we will denote by $\wp$ a prime of $\mathbb{Q}(\mu_{p-1})$ lying above p. Note that $\wp$ is unramified. We continue to denote the Teichmuller character by ω.

Lemma 3.1. *Let k be even and $2 \le k \le p-3$. Then the q-expansions of the modular forms $G_{2,\omega^{k-2}}$ and $G_{1,\omega^{k-1}}$ have $\wp$-integral coefficients in $\mathbb{Q}(\mu_{p-1})$ and are congruent modulo $\wp$ to the q-expansion*

$$-\frac{B_k}{2k} + \sum_{n\ge 1}\sum_{d|n} d^{k-1}q^n.$$

Proof. Since $\omega(d) \equiv d \bmod \wp$, $\omega^{k-2}(d)d \equiv d^{k-1} \bmod \wp$ and $\omega^{k-1}(d) \equiv d^{k-1} \bmod \mathfrak{p}$. Hence it suffices to investigate the constant terms only. We know that (see (6) and (7) of Appendix)

$$L(0,\epsilon) = \frac{-1}{p}\sum_{n=1}^{p-1}\epsilon(n)\left(n - \frac{p}{2}\right),$$

$$L(-1,\epsilon) = \frac{-1}{2p}\sum_{n=1}^{p-1}\epsilon(n)\left(n^2 - pn - \frac{p^2}{6}\right).$$

Since we know that $\omega(n) \equiv n^p \bmod (\wp^2)$ (cf section 6.5 of Appendix), we find that

$$pL(0,\omega^{k-1}) \equiv -\sum_{n=1}^{p-1} n^{1+p(k-1)} \bmod \wp^2,$$

$$pL(-1,\omega^{k-2}) \equiv -\frac{1}{2}\sum_{n=1}^{p-1} n^{2+p(k-2)} \bmod \wp^2.$$

Note that $\sum_{n=1}^{p-1}\epsilon(n)n \equiv 0 \bmod \wp$ when ϵ is an even character. Moreover, we know that (see proposition 6.6 of Appendix)

$$pB_t \equiv \sum_{n=1}^{p-1} n^t \bmod p^2.$$

Therefore, we have

$$L(0,\omega^{k-1}) \equiv -\frac{1}{2}B_{1+p(k-1)} \equiv -\frac{1}{2}(1+p(k-1))\frac{B_k}{k} \equiv -\frac{B_k}{k} \bmod \wp,$$

$$L(-1,\omega^{k-2}) \equiv -\frac{1}{2}B_{2+p(k-2)} \equiv -\frac{1}{2}(2+p(k-2))\frac{B_k}{k} \equiv -\frac{B_k}{k} \bmod \wp.$$

For the second equivalence of each statement above, we use Kummer congruence as explained in proposition 6.4 in the Appendix. Note that

$$1 + p(k - 1) = k + (p - 1)(k - 1) \equiv k \bmod (p - 1),$$

$$2 + p(k - 2) = k + (p - 1)(k - 2) \equiv k \bmod (p - 1). \qquad \square$$

The following corollary is now obvious.

Corollary 3.2. *Let k be even and $2 \leq k \leq p - 3$. Let n, m be even integers such that $n + m \equiv k \bmod (p - 1)$ and $2 \leq n, m \leq p - 3$. The the product $G_{1,\omega^{n-1}} G_{1,\omega^{m-1}}$ is a modular form of weight 2 and type ω^{k-2} whose q-expansion coefficients are $\wp$-integral in $\mathbb{Q}(\mu_{p-1})$. Its constant term is a $\wp$-adic unit if neither B_n nor B_m is divisible by p.*

The next theorem guarantees the existence of the Eisenstein series we are looking for.

Theorem 3.3. *Let k be an even integer $2 \leq k \leq p - 3$. Then there exists a modular form g of weight 2 and type ω^{k-2} whose q-expansion coefficients are $\wp$-integers in $\mathbb{Q}(\mu_{p-1})$ and whose constant term is a $\wp$-unit.*

Proof.

Case (i). If $p \nmid B_k$, we can take $G_{2,\omega^{k-2}}$ by lemma 3.1.

Case (ii). If we have a pair of even integers m n such that $n + m \equiv k \bmod (p - 1)$, $2 \leq n, m \leq p - 3$ and $p \nmid B_m B_n$, then we can take $G_{1,\omega^{n-1}} G_{1,\omega^{m-1}}$ by corollary 3.2.

Case (iii). Suppose neither of the above two cases are true. We will show that consequently too many Bernoulli numbers will be p-divisible, which will lead to violation of an upper bound for the p-part h_p^* of the relative class number of $\mathbb{Q}(\mu_p)$. Let t be the number of even integers n, $2 \leq n \leq p - 3$ such that p divides B_n. It is easy to see that $t \geq \frac{p-1}{4}$ if the cases (i) and (ii) do not arise. But then, p^t must divide h_p^* (see section 6.2 of Appendix). However, that contradicts a result of Carlitz, which says that $h_p^* < p^{(\frac{p-1}{4})}$. Hence we must be in either in case (i) or case (ii). $\square$

4. Existence of a semi-cusp form with suitable eigenvalues

In this section, we will first construct a semi-cusp form f which is a simultaneous eigenvector modulo $\wp$ for all Hecke operators T_l with $(p, l) = 1$. Then we will lift f to a semi-cusp form f' which is an eigenvector for all such T_l's.

Fix an even integer k, $2 \leq k \leq p - 3$ and assume that $p | B_k$. Consider $\epsilon = \omega^{k-2}$. Since $B_2 = \frac{1}{6}$, k is at least 4, and hence ϵ is a non-trivial even character. We will only be interested in modular forms of weight 2 and type ϵ.

Proposition 4.1. *There exists a semi-cusp form $f = \sum_{n \geq 1} a_n q^n$ such that a_n are $\wp$-integers in $\mathbb{Q}(\mu_{p-1})$ and such that $f \equiv G_{2,\epsilon} \equiv G_k \bmod \wp$.*

Proof. Consider $f = G_{2,\epsilon} - c.g$, where c is the constant term of $G_{2,\epsilon}$. Then f is a semi-cusp form. Now, $c \in \wp$ as $p | B_k$. Hence, $f \equiv G_{2,\epsilon} \equiv G_k \bmod \wp$. $\square$

Observe further that f is a mod $\wp$-eigenform for all Hecke operators T_l with $(l, p) = 1$, as the Eisenstein series $G_{2,\epsilon}$ is an eigenform form for all such T_l with eigenvalue $(1 + \epsilon(l)l)$. Therefore,

$$T_l(f) \equiv T_l(G_{2,\epsilon}) \equiv (1 + \epsilon(l)l)G_{2,\epsilon} \equiv (1 + \epsilon(l)l)f \text{ modulo } \wp. \tag{5}$$

4.1 *Deligne–Serre lifting lemma*

The following result of Deligne and Serre [D-S] ensures that there exists a semi-cusp form f' which is an eigenvector for the T_l's ($(l, p) = 1$) with eigenvalues congruent modulo $\wp$ to those of the mod-$\wp$ eigenvector f obtained previously.

Lemma 4.2. *Let M be a free module of finite rank over a discrete valuation ring R with residue field k, fraction field K and maximal ideal $\mathfrak{m}$. Let S be a (possibly infinite) set of commuting R-endomorphisms of M. Let $0 \neq f \in M$ be an eigenvector modulo $\mathfrak{m}M$ for all operators in S, i.e., $Tf = a_T f$ mod $\mathfrak{m}M$ $\forall T \in S$ ($a_T \in R$). Then there exists a DVR R' containing R with maximal ideal $\mathfrak{m}'$ containing $\mathfrak{m}$, whose field of fractions K' is a finite extension of K and a non-zero vector $f' \in R' \otimes_R M$ such that $Tf' = a'_T f'$ for all $T \in S$ with eigenvalues a'_T satisfying $a'_T \equiv a_T$ mod $\mathfrak{m}'$.*

Proof. Let $\mathbb{T}$ be the algebra generated by S over R. Clearly $\mathbb{T} \in End_R(M)$. As M is an free R-module of finite rank, so is $End_R(M)$. Therefore, $\mathbb{T}$ is also free module of finite rank over R, generated by $T_1, \ldots, T_r \in S$. Let h_i denote the minimal polynomial of T_i acting on $K \otimes_R M$. If we adjoin the roots of all such minimal polynomials to K, we get a finite extension K' of K. The integral closure of R in K' gives us a DVR R' with maximal ideal $\mathfrak{m}'$ lying over m, and with residue field k' containing k. By replacing M with $R' \otimes M$ and $\mathbb{T}$ with $R' \otimes_R \mathbb{T}$, we will continue to write R, $\mathfrak{m}$, k, K in stead of R', $\mathfrak{m}$ etc.

Consider the ring homomorphism $\lambda : \mathbb{T} \longrightarrow k$ given by $T \mapsto a_T$ mod $\mathfrak{m}$ for all T in S. Clearly, $ker(\lambda)$ is a maximal ideal of $\mathbb{T}$. Choose a minimal prime $\wp$ in $ker(\lambda)$. Then, $\wp$ is contained in the set of zero-divisors of $\mathbb{T}$ (see proposition 6.9 of Appendix). As $\mathbb{T}$ is a free R-module, R contains no zero-divisors of $\mathbb{T}$ and hence, $\mathfrak{p} \cap R = \{0\}$. Thus, $\mathbb{T}/\mathfrak{p}$ is a finite integral extension of R. Let L denote the field of fractions of the integral domain $\mathbb{T}/\mathfrak{p}$. Let R_L be the integral closure of R in L, then R_L is a DVR with maximal ideal m_L containing $\mathfrak{m}$ and residue field l containing k.

Consider the map $\lambda' : \mathbb{T} \longrightarrow \mathbb{T}/\mathfrak{p} (\hookrightarrow R_L)$ given by reduction modulo $\mathfrak{p}$. Let $\lambda'(T) = a'_T$ for all $T \in S$. Clearly, λ' maps the maximal ideal $ker(\lambda)$ of $\mathbb{T}$ into the maximal ideal m_L of R_L. But $(T - a_T) \in ker(\lambda)$, hence $\lambda'(T - a_T) \in m_L$ i.e., $a'_T \equiv a_T$ modulo m_L.

Now consider the ring $K \otimes_R \mathbb{T}$. It is an Artinian ring, hence it has finitely many maximal ideals with residue fields all isomorphic to K. Let $\mathcal{P}$ be the prime ideal in $K \otimes \mathbb{T}$ generated by $\mathfrak{p}$. It will suffice to show that $\mathcal{P}$ is an associated prime of $K \otimes M$. Note that $\wp \subset ker(\lambda)$ implies $\wp$ annihilates f in $M/\mathfrak{m}$. Now let $x \in Ann_{\mathbb{T}/\mathfrak{m}}(f)$, say $x = \bar{g}(T_1, \ldots, T_n)$. Then, $x = \bar{g}(a'_{T_1}, \ldots, a'_{T_1})$ modulo $(T_1 - a'_{T_1}, \ldots, T_n - a'_{T_n})$. Thus, $xf = \bar{g}(a'_{T_1}, \ldots, a'_{T_1})f$ modulo $m_L M$, noting that $T - a'_T \in \wp$, and $\wp$ annihilates f modulo $m_L M$. As $a'_T \equiv a_T$ mod m_L, we must have $\bar{g}(a_{T_1}, \ldots, a_{T_1})f = 0$ mod

$m_L M$. As $f \neq 0$, we must have $\bar{g}(a_{T_1}, \ldots, a_{T_1}) = 0$ in l. Thus, $x \in \wp$, and $\wp = Ann_{\mathbb{T}/\mathfrak{m}}(f)$ is an associated prime of $M/\mathfrak{m}$. For proof of the following two statements, see section 6.8.2 of Appendix.

(i) $\mathfrak{p}$ is in $Assoc_{\mathbb{T}/m}(M/\mathfrak{m})$, hence in $Supp_{\mathbb{T}/m}(M/\mathfrak{m})$, and hence $Ann_{\mathbb{T}/m}(M/\mathfrak{m}) \subset \wp$.
(ii) Now, it follows that $Ann_{K\otimes\mathbb{T}}(K \otimes M) \subset \mathcal{P}$, hence $\mathcal{P} \in Supp_{K\otimes\mathbb{T}}(K \otimes M)$ and therefore $\mathcal{P}$ is in $Assoc_{K\otimes\mathbb{T}}(K \otimes M)$.

Now, $\mathcal{P}$ is the annihilator of some $0 \neq f'' \in K \otimes M$, hence $\mathcal{P}$ annihilates some $f' \in M$. As $T - a'_T \in \mathfrak{p}$, we have $T - a'_T \in \mathcal{P}$ and $(T - a'_T)(f') = 0$. Thus, $Tf' = a'_T f'$ where $a'_T \equiv a_T$ modulo m_L, which concludes our proof. □

4.2 *Lifting the semi-cusp form to an eigenvector for T_n for $(n, p) = 1$*

The following theorem ensures that we have a semi-cusp form which is an eigenvector for all Hecke operators T_n with $p \nmid n$.

Theorem 4.3. *There is a semi-cusp form $f' = \sum_{n=1}^{\infty} c_n q^n$ of weight 2 and type ϵ such that all its coefficients are defined over a finite extension of L of $\mathbb{Q}(\mu_{p-1})$ and are $\wp_L$-integral where $\wp_L$ is a prime above p. Further, $T_l f' \equiv (1 + \epsilon(l)l) f'$ modulo $\wp_L$.*

Proof. There is a basis B of $S'_2(\Gamma_1(p), \epsilon)$ consisting of semi-cusp forms all of whose coefficients are defined over a finite extension K of $\mathbb{Q}(\mu_{p-1})$. Let R be the localization of the ring of integers of K at a prime $\wp_K$ above $\wp$. Let M be the free R-module of semi-cusp forms generated by B. Let $S = \{T_n | (p, n) = 1\}$. We know by proposition 4.1 and (5) that there exists $f \in M$ such that

$$T_l(f) \equiv (1 + \epsilon(l)l) f \text{ modulo } \wp.$$

By applying the lifting lemma 4.2, we can conclude that there is a finite extension L of K with a prime $\wp_L$ over $\wp_K$ such that there exists a semi-cusp form f', with $\wp_L$-integral coefficients in L such that $T_l(f') = c_l f'$ and $c_l \equiv 1 + \epsilon(l)l$ modulo $\wp_L$. □

5. Construction of cusp eigenform

We will first show that the semi-cusp form f' obtained in the previous section is in fact a cusp form. Then, we will finally show that the cusp form f' must be an eigenvector for all Hecke operators T_n including those n which are not co-prime to p.

5.1 *Existence of a suitable cusp form*

Proposition 5.1. *There exists a non-zero cusp form f' of type ϵ, which is an eigenform for all Hecke operators T_n with $(n, p) = 1$, and which has the property that for any prime $l \neq p$, the eigenvalue λ_l of T_l acting on f' satisfies*

$$\lambda_l \equiv 1 + l^{k-1} \equiv 1 + \epsilon(l)l \ mod\ \wp_L,$$

where $\wp_L$ is a certain prime (independent of l) lying over $\wp$ in the field $L = \mathbb{Q}(\mu_{p-1}, \lambda_n)$ generated by the eigenvalues over $\mathbb{Q}(\mu_{p-1})$.

Proof. We already established the existence of a semi-cusp form f' which is an eigenform for all Hecke operators T_n $(n, p) = 1$ whose eigenvalues have the required congruence properties. It suffices to assert that f' is in fact a cusp form. As $M_2(\Gamma_0(p), \epsilon)$ is spanned by the cusp forms, the semi-cusp form $S_{2,\epsilon}$ and the Eisenstein series $G_{2,\epsilon}$, we must have

$$S_2'(\Gamma_1(p), \epsilon) = S_2(\Gamma_1(p), \epsilon) \oplus \mathbb{C}s_{2,\epsilon},$$

where orthogonality of the Eisenstein space and the space of cusp forms under Petersson inner product $\langle , \rangle$ is the reason behind the above sum being a direct one (see section 6.6 of Appendix). Suppose $f' = h + as_{2,\epsilon}$ $(a \neq 0)$. Then, $f' - as_{2,\epsilon} \in S_2(\Gamma_1(p), \epsilon)$. But, $f' - as_{2,\epsilon} \in \mathcal{E}_2(\Gamma_1(p), \epsilon)$ as well, where $\mathcal{E}_2(\Gamma_1(p), \epsilon)$ denotes the subspace consisting of Eisenstein series in $M_2(\Gamma_1(p), \epsilon)$. As the orthogonal subspaces $\mathcal{E}_2(\Gamma_1(p), \epsilon)$ and $S_2(\Gamma_1(p), \epsilon)$ have trivial intersection, $f' - as_{2,\epsilon} = 0$, i.e., $f' = as_{2,\epsilon}$. Applying T_l to both sides, $(l \neq p)$, we see that we must have $1 + \epsilon(l)l \equiv l + \epsilon(l) \bmod \wp_L$, which forces $\epsilon(l) = 1$. But ϵ is a non-trivial character and $l \neq p$ is arbitrary, hence f' must be a cusp form. □

5.2 *Operators T_n for $(n, p) \neq 1$*

So far, we know that we have a cusp form f for $\Gamma_1(p)$ of weight 2 and type ϵ which is an eigenform for all Hecke operators T_l $(l, p) = 1$. In this section we will assert that f is in fact a common eigenform for all Hecke operators, including T_n $(n, p) \neq 1$.

Proposition 5.2. *Any form f' as above is an eigenform for all Hecke operators (including those for which $p|n$). Hence, after replacing f' by a suitable multiple of f', we have*

$$f' = \sum_{n=1}^{\infty} \lambda_n q^n, \qquad \textit{where} \quad T_n(f') = \lambda_n f'.$$

Proof. f' must be a newform. For, if it were an old form it will have to originate from a non-zero modular form in $M_2(SL_2(\mathbb{Z}))$, but that space is trivial. Now for a new form f', if it is an eigenform for T_n $((n, p) = 1)$ it has to be an eigenform for all T_n by the theory of newforms (see Theorem 5.8.2 of [Di-S]). Now we can take a suitable multiple of f' to get a normalized cusp eigenform as prescribed in the theorem. □

Remark. The cusp eigenform obtained above can be associated to a Galois representation which finally gives an unramified p-extension of $\mathbb{Q}(\mu_p)$, where μ_p denotes the p-power roots of unity for an odd prime p. This exposition can be found in the article by C. S. Dalawat [D] in this volume.

6. Appendix

Here we provide a brief discussion of the various ingredients used in the previous sections.

6.1 *Dirichlet L-functions*

A Dirichlet character is a homomorphism $\chi : (\frac{\mathbb{Z}}{N\mathbb{Z}})^{\times} \longrightarrow \mathbb{C}^{\times}$, where N is any positive integer, and $A^{\times}$ denote the multiplicative group of units in a ring A. N is called the conductor of χ if χ does not factor through $(\frac{\mathbb{Z}}{M\mathbb{Z}})^{\times}$ for any $M < N$. We denote the conductor of χ by f_{χ}. We can easily extend the definition of χ to $\mathbb{Z}$ by setting $\chi(n) = \chi(n \bmod N)$ if $(n, N) = 1$ and $\chi(n) = 0$ otherwise. The Dirichlet L-series of χ is defined as

$$L(s, \chi) = \sum_{n=1}^{\infty} \chi(n) n^{-s},$$

where s is a complex number with $Re(s) > 1$. It is well-known that $L(s, \chi)$ can be analytically continued to the whole complex plane except a simple pole of residue 1 at $s = 1$ when χ is the trivial character (in which case the function is just the Riemann-zeta function). Further, $L(s, \chi)$ satisfies a functional equation relating its values at $s = 1$ to values $1 - s$. It also has a Euler product, i.e.,

$$L(s, \chi) = \prod_{l} (1 - \chi(l) l^{-s})^{-1}, \quad Re(s) > 1$$

where l runs over the rational primes. The Dirichlet L-functions are related to the Dedekind zeta function of an abelian number field, as explained below.

Recall that for a number field K, the Dedekind zeta function is defined as

$$\zeta_K(s) = \sum_{\mathfrak{a}} (N\mathfrak{a})^{-s}, \qquad Re(s) > 1,$$

where $\mathfrak{a}$ runs over the ideals of the ring $\mathcal{O}_K$ of integers in K. It is well-known that $\zeta_K(s)$ can be analytically continued to the whole complex plane except for a simple pole at $s = 1$. Further, $\zeta_K(s)$ satisfies a functional equation, relating the values at s to values at $1 - s$.

We can view χ as a Galois character

$$\chi \, : \, Gal(\mathbb{Q}(\mu_N)/\mathbb{Q}) \simeq (\mathbb{Z}/N\mathbb{Z})^{\times} \longrightarrow \mathbb{C}^{\times},$$

and this gives a correspondence $\chi \rightarrow$ fixed subfield of $ker(\chi)$ in $\mathbb{Q}(\mu_N)$, which is an abelian extension of $\mathbb{Q}$. This leads to a one-to-one correspondence between groups of Dirichlet characters and abelian extensions of $\mathbb{Q}$. If K is an abelian extension of $\mathbb{Q}$, it is contained in some $\mathbb{Q}(\mu_N)$ and there will be a corresponding group X of Dirichlet characters of conductor dividing N.

If K is an abelian number field and X is the corresponding group of Dirichlet characters, then one can show that (see theorem 4.3 in [Wa])

$$\zeta_K(s) = \prod_{\chi \in X} L(s, \chi).$$

6.2 *The relative class number and Dirichlet L-values*

The analytic class number formula is given by

$$\lim_{s\to 1}\zeta_K(s)=\frac{2^{r_K}(2\pi)^{t_K}h_K R_K}{w_K\sqrt{|d_K|}},$$

where r_K and t_K denote respectively the number of real and complex pairs of embedding of K, w_K the number of roots of unity in K, R_K the regulator of K, d_K the discriminant of K and h_K the class number of K.

Now consider $K=\mathbb{Q}(\zeta_p)$, then $r_K=0$, $t_K=\frac{p-1}{2}$. Let K^+ be the maximal real subfield of K, for which $r_{K^+}=\frac{p-1}{2}$ and $t_{K^+}=0$. It is easy to establish that h_{K^+} divides h_K. The relative class number of K is defined as $h_K^-=\frac{h_K}{h_{K^+}}$. The purpose of this section is to investigate the p-part h_K^-, and relate it to the values of Dirichlet L-functions.

Proposition 6.1.

$$h_K^-=\alpha p\prod_{i=0}^{p-2}L(0,w^i),$$

where α is a certain power of 2.

Proof. Dividing the analytic class number formulas for K and K^+, and then shifting the limit to $s\to 0$ via the functional equations, one can cancel out the extraneous factors and deduce that (see [Gr])

$$h_K^-=\frac{w_K}{2^e w_{K^+}}\lim_{s\to 0}\frac{\zeta_K(s)}{\zeta_{K^+}(s)},$$

where $\frac{R_K}{R_{K^+}}=2^e$. But

$$\zeta_K(s)=\prod_{i=0}^{p-2}L(0,w^i),\quad \zeta_{K^+}(s)=\prod_{i\text{ even}}^{p-2}L(0,w^i).$$

Now observing that $w_K=2p$ and $w_{K^+}=2$, we obtain the desired result. □

6.3 *Dirichlet L-values and Bernoulli numbers*

Recall that Bernoulli numbers B_n are given by

$$\frac{t}{e^t-1}=\sum_{n=0}^{\infty}B_n\frac{t^n}{n!}.$$

Eg, $B_0=1$, $B_1=-\frac{1}{2}$, $B_2=\frac{1}{6}$ etc.

The n-th Bernoulli polynomial $B_n(X)$ is defined by

$$\frac{te^{Xt}}{e^{tX}-1} = \sum_{n=0}^{\infty} B_n(X)\frac{t^n}{n!}.$$

It is easy to see that

$$B_n(X) = \sum_{i=o}^{n} \binom{n}{i} B_i X^{n-i}.$$

Eg, $B_1(X) = X - \frac{1}{2}$, $B_2(X) = X^2 - X + \frac{1}{6}$, etc.

Now, for a Dirichlet character χ of conductor f, we define the generalized Bernoulli numbers $B_{n,\chi}$ by

$$\sum_{a=1}^{f} \frac{\chi(a)te^{at}}{e^{ft}-1} = \sum_{n=0}^{\infty} B_{n,\chi}\frac{t^n}{n!}.$$

The following well-known proposition allows us to express generalized Bernoulli numbers in terms of Bernoulli polynomials (cf [Wa]).

Proposition 6.2. *If g is any multiple of f, then*

$$B_{n,\chi} = g^{n-1}\sum_{a=1}^{g} \chi(a) B_n\left(\frac{a}{g}\right).$$

Proof.

$$\begin{aligned}
&\sum_{n=0}^{\infty} g^{n-1}\sum_{a=1}^{g} \chi(a) B_n\left(\frac{a}{g}\right)\frac{t^n}{n!} \\
&\quad = \sum_{a=1}^{g} \chi(a)\frac{1}{g}\frac{(gt)e^{(\frac{a}{g})gt}}{e^{gt}-1} \\
&\quad = \sum_{b=1}^{f}\sum_{c=0}^{h-1} \chi(b+cf)\frac{te^{(b+cf)t}}{e^{fht}-1} \quad \text{where } g = hf, \quad a = b + cf \\
&\quad = \sum_{b=1}^{f} \frac{\chi(b)te^{bt}}{e^{ft}-1} \\
&\quad = \sum_{n=0}^{\infty} B_{n,\chi}\frac{t^n}{n!}.
\end{aligned}$$

□

For example,

$$B_{1,\chi} = \sum_{a=1}^{f} \chi(a)\left(\frac{a}{f} - \frac{1}{2}\right) = \frac{1}{f}\sum_{a=1}^{f} \chi(a)\left(a - \frac{1}{2}f\right).$$

$$B_{2,\chi} = f\sum_{a=1}^{f} \chi(a)\left(\left(\frac{a}{f}\right)^2 - \frac{1}{2}\frac{a}{f} + \frac{1}{6}\right) = \frac{1}{f}\sum_{a=1}^{f} \chi(a)\left(a^2 - fa + \frac{f^2}{6}\right).$$

The generalized Bernoulli numbers can be related to the values of Dirichlet L-values as follows:

Proposition 6.3. $L(1-n,\chi) = -\frac{B_{n,\chi}}{n}$, $n \geq 1$.

For example, if χ is a Dirichlet character modulo p, we have

$$L(0,\chi) = -B_{1,\chi} = -\frac{1}{p}\sum_{n=1}^{p} \chi(n)\left(n - \frac{1}{2}p\right). \tag{6}$$

$$L(-1,\chi) = -B_{2,\chi} = -\frac{1}{2p}\sum_{n=1}^{p} \chi(a)\left(n^2 - pn + \frac{p^2}{6}\right). \tag{7}$$

6.4 *Some congruences involving Bernoulli numbers*

We require the following congruences involving Bernoulli numbers.

Proposition 6.4 (Kummer Congruence). $\frac{B_m}{m} \equiv \frac{B_n}{n}$ *if* $m \equiv n \not\equiv 0$ *mod* $(p-1)$.

Kummer's congruence can be proved in the following manner (cf. [B-S]): let g be a primitive root mod p. Consider

$$F(t) = \frac{gt}{e^{gt}-1} - \frac{t}{e^t - 1} = \sum_{m=1}^{\infty} (g^m - 1)B_m \frac{t^m}{m!}. \tag{8}$$

Letting $e^t - 1 = u$, we can write

$$F(t) = \frac{gt}{(1+u)^g - 1} - \frac{t}{u} = tG(u), \quad \text{where}$$

$$G(u) = \frac{g}{(1+u)^g - 1} - \frac{1}{u} = \sum_{k=1}^{\infty} c_k u^k, \quad c_k \in \mathbb{Z}.$$

Now,

$$G(u) = G(e^t - 1) = \sum_{k=0}^{\infty} c_k (e^t - 1)^k = \sum_{m=1}^{\infty} A_m \frac{t^m}{m!}. \tag{9}$$

But A_m are p-integral as they are integral linear combinations of c_k's. Further, they have period $(p-1)$ modulo p, as the coefficients r^n of $\frac{t^n}{n!}$ in e^{rt} $(r \geq 0)$ have that periodicity by Fermat's little theorem $r^{n+p-1} \equiv r^n$ modulo p. Comparing coefficients in (8) and (9), we obtain

$$\frac{g^m - 1}{m!} B_m = \frac{A_{m-1}}{(m-1)!} \Rightarrow \frac{B_m}{m}(g^m - 1) = A_{m-1}.$$

If $p - 1 \not| m$, then $g^m - 1 \not\equiv 0 \bmod p$ as g is a primitive root mod p. Clearly, $g^m - 1$ has period $p - 1$ mod p. Therefore, $\frac{B_m}{m}$ also has period $p - 1$ mod p and is p-integral.

Proposition 6.5. *pB_m is p-integral, and B_m is p-integral if $(p-1) \not| m$.*

Proposition 6.6. *For an even integer m, $pB_m \equiv \sum_{a=1}^{p-1} a^m$ modulo p^2 if $p \geq 5$.*

We can easily prove the above two propositions using the following lemma.

Lemma 6.7. $(m+1)S_m(n) = \sum_{k=0}^{m} \binom{m+1}{k} B_k n^{m+1-k}$, *where* $S_m(n) = 1^n + 2^n + \cdots + m^n$.

Proof.

$$\sum_{m=0}^{\infty} S_m(n)\frac{t^m}{m!} = \sum_{a=0}^{n-1} \frac{e^{nt}-1}{e^t-1} = \frac{e^{nt}-1}{t}\frac{t}{e^t-1} = \sum_{l=1}^{\infty} n^l \frac{t^{l-1}}{l!} \sum_{k=0}^{\infty} B_k \frac{t^k}{k!}$$

$$\Rightarrow \frac{S_m(n)}{m!} = \sum_{k=0}^{m+1} \frac{B_k}{(m+1-k)!k!} n^{m+1-k}$$

$$\Rightarrow (m+1)!\frac{S_m(n)}{m!} = \sum_{k=0}^{m+1} \binom{m+1}{k} B_k n^{m+1-k}$$

□

In order to prove proposition 6.5, it is enough to show that $pB_m \equiv S_m(p)$ modulo p. It is clear that $S_m(p) \equiv 0 \bmod p$ if $(p-1) \not| m$ and $S_m(p) \equiv p - 1 \bmod p$ if $(p-1)|m$. By our lemma, we have

$$S_m(p) = pB_m + \binom{m}{1} B_{m-1}\frac{p^2}{2} + \binom{m}{2} B_{m-2}\frac{p^3}{3} + \cdots + \binom{m}{m} B_0 \frac{p^{k+1}}{k+1}. \quad (10)$$

Clearly, $\frac{p^{k+1}}{k+1} \equiv 0 \bmod p$ for $k \geq 2$, and $\frac{p^{k+1}}{k+1}$ is p-integral even for $k = 1$. Applying induction, let pB_j be p-integral for $j < m$. Then, pB_m is p-integral as well, and we also obtain $S_m(n) \equiv pB_m \bmod p$ from (10). Note that though we need the result only for odd prime p, the above proof works for $p = 2$ as well, as B_n vanishes for odd $n \geq 3$.

To prove proposition 6.6, it suffices to establish that $ord_p(\binom{m}{k} B_{m-k}\frac{p^{k+1}}{k+1}) \geq 2$ in view of (10). Since pB_{m-k} is p-integral, we need only $k - ord_p(k+1) \geq 2$. For $p \geq 5$ and $k \geq 2$, it is obvious. For $k = 1$, note that $B_{m-1} = 0$ unless $m = 2$, which again follows trivially.

6.5 *A refined congruence for the Teichmuller character*

Let $w : (\mathbb{Z}/p\mathbb{Z})^{\times} \longrightarrow \mu_{p-1}$ be the character given by $w(n) \equiv n$ modulo $\wp$ where $\wp$ is any prime ideal above p in $\mathbb{Q}(\mu_{p-1})$. The character w is known as the Teichmuller character. We have used the following congruence for the Teichmuller character.

Proposition 6.8. *For $(n, p) = 1$, we have $w(n) \equiv n^p$ modulo $\wp^2$ where $\wp$ is a fixed prime above p in $K = \mathbb{Q}(\mu_{p-1})$.*

Proof. Let us recall Hensel's lemma:

Let R be a ring which is complete with respect to an ideal I and let $f(x) \in R[x]$. If $f(a) \equiv 0 \bmod (f'(a)^2 I)$ then there exists $b \in R$ with $b \equiv a$ modulo $(f'(a)I)$ such that $f(b) = 0$. Further, b is unique if $f'(a)$ is a non-zero divisor in R.

Now let $K_{\wp}$ be the completion of K at $\wp$. Let $R = \mathcal{O}_{\wp}$ be the completion of the ring of integers $\mathcal{O}$ of K with respect to $\wp$. Let $I = \wp^2$, then we can also think of R as the completion of $\mathcal{O}$ with respect to I. Consider $f(x) = x^{p-1} - 1$ and let $a = n^p$, where $(n, p) = 1$. Then,

$$f(a) = (n^p)^{p-1} - 1 \equiv 0 \bmod \wp^2, \text{ as } \#\left(\frac{\mathcal{O}_{\wp}}{\wp^2}\right)^{\times}$$

$$= \#\left(\frac{\mathcal{O}}{\wp^2}\right)^{\times} = N\wp^2 - N\wp = p(p-1).$$

Moreover $f'(a) = (p-1)a^{p-2}$ is not a zero-divisor in R. Therefore by Hensel's lemma there exists a unique b_n in R such that $b_n^{p-1} - 1 = 0$ and $b_n \equiv n^p$ modulo $\wp^2$. Now, if we define $w(n) = b_n$, we obtain the Teichmuller character $w : \left(\frac{\mathbb{Z}}{p\mathbb{Z}}\right)^{\times} \longrightarrow \mu_{p-1}$ with the more refined congruence $w(n) \equiv n^p$ modulo $\wp^2$. □

6.6 *Petersson inner product*

There is a measure on the upper half complex plane $\mathfrak{h}$ given by $d\mu(\tau) = \frac{dx\,dy}{y^2}$ where $\tau = x + iy \in \mathfrak{h}$. It is easy to show that $d\mu(\tau)$ is invariant under $GL_2(\mathbb{R})^+ \subset Aut(\mathfrak{h})$, i.e., $d\mu(\alpha\tau) = d\mu(\tau)$. In particular, the measure is $SL_2(\mathbb{Z})$-invariant. As $\mathbb{Q} \cup \{\infty\}$ is a countable set of measure 0, $d\mu$ suffices for integration over the extended upper half plane $\mathfrak{h}^* = \mathfrak{h} \cup \mathbb{Q} \cup \{\infty\}$. Let D^* be the fundamental domain for $SL_2(\mathbb{Z})$, i.e.,

$$D^* = \mathfrak{h}^*/SL_2(\mathbb{Z}) = \left\{\tau \in \mathfrak{h} | Re(\tau) \leq \frac{1}{2}, \quad |\tau| \geq 1\right\} \cup \{\infty\}.$$

For a congruence subgroup Γ of $SL_2(\mathbb{Z})$, we have $(\pm I)\Gamma\ SL_2(Z) = \bigcup_j (\pm 1)\Gamma\alpha_j$ where j runs over a finite set. Then, the fundamental domain for Γ is given by

$$X(\Gamma) = \mathfrak{h}^*/\Gamma = \bigcup \alpha_j(D^*).$$

This allows us to integrate function of $\mathfrak{h}^*$ invariant under Γ by setting

$$\int_{X(\Gamma)} \phi(\tau) d\mu(\tau) = \int_{\bigcup_j \alpha_j(D^*)} \phi(\tau) d\mu(\tau) = \sum_j \int_{D^*} \phi(\alpha_j(\tau)) d\mu(\tau).$$

By letting $V_\Gamma = \int_{X(\Gamma)} d\mu(\tau)$, we can define an inner product

$$\langle\, ,\rangle_\Gamma \;:\; S_k(\Gamma) \times M_k(\Gamma) \longrightarrow \mathbb{C}.$$

given by

$$\langle f, g\rangle_\Gamma = \frac{1}{V_\Gamma} \int_{X(\Gamma)} f(\tau)\overline{g(\tau)}(Im(\tau))^k d\mu(\tau).$$

Note that the integrand is invariant under Γ. For the integral to converge, we need one of f or g to be a cusp form (see section 5.4 in [Di-S]). Clearly this inner product is Hermitian and positive definite. When we take a modular form $f \in M_k(\Gamma) - S_k(\Gamma)$, we can show that f is orthogonal under $\langle\, ,\rangle_\Gamma$ to all of $S_k(\Gamma)$. Thus, we can think of the quotient space $\mathcal{E}_k(\Gamma) = M_k(\Gamma)/S_k(\Gamma)$ as the complementary subspace linearly disjoint from $S_k(\Gamma)$. This allows us to write

$$M_k(\Gamma) = S_k(\Gamma) \oplus \mathcal{E}_k(\Gamma).$$

6.7 *Hecke operators*

For any $\alpha \in GL_2(\mathbb{Q})$, one can write the double coset $\Gamma\alpha\Gamma = \bigcup_i \Gamma\alpha_i$ where α_i runs over a finite set. We can define an action of the double coset on $M_k(\Gamma)$ by setting $f|\Gamma\alpha\Gamma = \sum f|[\alpha_i]$. It is easy to verify that these operators preserve $M_k(\Gamma)$, $S_k(\Gamma)$ and $\mathcal{E}_k(\Gamma)$.

We need to consider only the case $\Gamma = \Gamma_1(p)$. For any integer d such that $(d, p) = 1$, we can define an operator $\langle d\rangle$ as follows: we have $ad - bp = 1$ for some $a, b \in \mathbb{Z}$. Taking $\alpha = \begin{bmatrix} a & b \\ p & d \end{bmatrix} \in \Gamma_0(p)$, we obtain

$$\langle d\rangle : M_k(\Gamma_1(p)) \longrightarrow M_k(\Gamma_1(p)),$$

$$\langle d\rangle f := f|\Gamma_1(p)\alpha\Gamma_1(p) = f|[\alpha]_k,$$

noting that $\Gamma_1(p)\alpha\Gamma_1(p) = \Gamma_1(p)\alpha$ as $\Gamma_1(p)$ is a normal subgroup of $\Gamma_0(p)$. The operators $\langle d\rangle$ are called diamond operators.

By taking $\alpha_l = \begin{bmatrix} 1 & 0 \\ 0 & l \end{bmatrix}$ for any prime l, we get an operator $T_l = f|\Gamma\alpha_l\Gamma$ for any prime l. We extend the definition of Hecke operators to all natural numbers inductively by setting

$$T_{l^{r+1}} = T_l T_{l^r} - l^{k-1}\langle l\rangle T_l^{r-1} \quad \text{for} \quad r \geq 1.$$

$$T_{mn} = T_m T_n \quad \text{when} \quad gcd(m, n) = 1$$

All these Hecke operators defined above are self adjoint with respect to the Petersson inner product. For more details, see chapter 5 of [Di-S]. A modular form is called an *eigenform* if it is a simultaneous eigenform for all Hecke operators T_n and $\langle d\rangle$, $(d, p) = 1$.

6.8 *Ingredients from commutative algebra*

The results proved below are required for the lifting lemma of Deligne and Serre in section 4.1.

6.8.1 *Minimal primes*

Let A be a commutative ring with 1. A prime ideal $\wp$ of A is called a *minimal prime* if it the smallest prime ideal (containing 0) in A. Such a prime exists by Zorn's lemma on the (non-empty as $1 \in A$) set S of primes ideals of A with the partial order $I \leq J$ when $J \subset I$, noting that any descending chain in S has its intersection as an upper bound in S.

Proposition 6.9. *A minimal prime $\wp$ of A is contained in the set Z of zero-divisors of A.*

Proof. Note that $x, y \in D = A - Z \Rightarrow xy \in D$. Thus D is a multiplicative set. On the other hand, $S = A - \wp$ is a maximal multiplicative closed set (as $\wp$ is a minimal prime). If $D \not\subset S$, then SD would be a multiplicative set strictly larger than S. Therefore, $D \subset S$ and $\wp \subset Z$. □

6.8.2 *Associated primes and support primes*

Let A be a commutative ring and M be an A-module. The annihilator of a submodule N of M is defined as

$$Ann_A(N) = \{a \in A | an = 0 \ \forall n \in N\}.$$

Clearly, $Ann_A(N)$ is an ideal of A. For an element $m \in M$, we can define its annihilator as $Ann_A(m) = \{a \in A | am = 0\}$.

Definition 6.10. *A prime ideal $\wp$ of A is called an* **associated prime** *if $\wp$ is the annihilator of some element of M. The set of associated primes of M in A is denoted by $Assoc_A(M)$.*

Proposition 6.11. *If M is non-zero and A is Noetherian, then $Assoc_A(M)$ is non-empty.*

Proof. Consider the set S of ideals ($\neq A$) of A which are annihilators of some element of M. As A is Noetherian, S has a maximal element, say $\wp$, which is necessarily the annihilator of some element m in M. Let $x, y \in A$ such that $xy \in \wp$ but $y \notin \wp$. Then $ym \neq 0$, but $\wp \subset (\wp, x) \subset Ann_A(ym) \in S$. It follows that $Ann_A(ym) = (\wp, x) = \wp$ by maximality of $\wp$. Therefore $x \in \wp$, and hence $\wp$ is an associated prime. □

Definition 6.12. *A prime ideal $\wp$ of A is called a* **support prime** *of M if $M_\wp \neq 0$.*

The set of support primes of M in A is denoted by $Supp_A(M)$.

Proposition 6.13. *Let A be Noetherian and M be a finitely generated A-module. Then $\wp \in Supp_A(M) \Leftrightarrow Ann_A(M) \subset \wp$*

Proof. Let $Ann_A(M) \not\subset \wp$. Then there exists $s \in A - \wp$ such that $sM = 0$, hence $M_\wp = 0$. Contra-positively, $\wp \in Supp_A(M)$ implies $Ann_A(M) \subset \wp$.

For the converse, let $m_1, \ldots, m_r$ generate M as an A-module. If $M_\wp = 0$, then we can find $s_i \in A - \wp$ such that $s_i m_i = 0$. Now $s = s_1 \ldots s_r \in A - \wp$ annihilates M, hence $Ann_A(M) \not\subset \wp$. □

Proposition 6.14. $Assoc_A(M) \subset Supp_A(M)$.

Proof. Let $\wp$ be an associated prime of M, say $\wp = Ann_A(m)$ for some $m \in M$. If $M_\wp = 0$ then there exists $s \in A - \wp$ such that $sm = 0$. But it would mean $s \in Ann_A(m) = \wp$, which is a contradiction. Thus, $M_\wp \neq 0$ and $\wp$ must be a support prime of M. □

Proposition 6.15. *Let A be a Noetherian ring and $\wp$ be a support prime. Then $\wp$ contains an associated prime $\mathfrak{q}$ of M.*

Proof. If $\wp$ is a support prime, $M_\wp \neq 0$. Then there must exist some $x \in M$ such that $(Ax)_\wp \neq 0$. Thus, there exists an associated prime $\mathfrak{q}$ of the A-module $(Ax)_\wp$. Hence there is an element $0 \neq \frac{y}{s}$ of $(Ax)_\wp$ with $y \in Ax$ and $s \notin \wp$ such that $\mathfrak{q}$ is the annihilator of $\frac{y}{s}$. Now, if there exists $b \in \mathfrak{q} - \wp$, then $b\frac{y}{s} = 0$ would imply $\frac{y}{s} = 0$, which is a contradiction.

Now we still have to show that $\mathfrak{q}$ is an associated prime of M as well. Let $b_1, \dots b_n$ be a set of generators of $\mathfrak{q}$. Then, there exists $t_i \in A - \wp$ such that $b_i t_i y = 0$. Let $t = t_1 . \dots . t_n$. Then, $\mathfrak{q}$ is the annihilator of $ty \in M$. □

Corollary 6.16. *If $\wp$ is a minimal prime in the support of M, then $\wp$ is also an associated prime when A is Noetherian.*

Proof. As $\wp$ must contain an associated prime, we get our result by minimality of $\wp$. □

Acknowledgement

I am very grateful to Kevin Buzzard for pointing out an error in an earlier version.

References

[BS] Z. I. Borevich and I. R. Shafarevich, Number Theory, Academic Press (1966).

[C] L. Carlitz, A generalization of Maillet's determinant and a bound for the first factor of the class number, *Proc. A.M.S.*, **12** (1961) 256–261.

[C-O] L. Carlitz and F. R. Olson, Maillet's determinant, *Proc. A.M.S.*, **6** (1955) 265–269.

[D] C. S. Dalawat, Ribet's modular construction of unramified p-extensions of $\mathbb{Q}(\mu_p)$ (to appear).

[D-S] P. Deligne and J-P. Serre, Formes modulaires de poids 1, *Ann. Scient. Ec. Norm. Sup., 4^e serie*, **7** (1974) 507–530.

[Di-S] F. Diamond and S. Shurman, A First Course on Modular Forms, Springer (2005).

[Gr] R. Greenberg, A generalization of Kummer's criterion, *Inventiones Math.*, **21** (1973) 247–254.

[La] S. Lang, Algebra, Springer-Verlag (2002).

[R] K. Ribet, A modular construction of unramified p-extensions of $\mathbb{Q}(\mu_p)$, *Inventiones Math.*, **34** (1976) 151–162.

[Wa] L. Washington, Introduction to Cyclotomic Fields, Springer-Verlag (1997).

Guwahati Workshop on Iwasawa Theory of Totally Real Fields No. 12, 2010, pp. 67–81.

Ribet's Modular Construction of Unramified p-extensions of $\mathbf{Q}(\mu_p)$

Chandan Singh Dalawat

Harish-Chandra Institute, Chhatnag Road, Jhunsi, Allahabad 211 019, India
e-mail: dalawat@gmail.com

Ribet's ingenious method is the gateway to much later progress.
— *Barry Mazur*

For every prime number p, we have the cyclotomic field $K = \mathbf{Q}(\zeta)$, obtained by adjoining to $\mathbf{Q}$ a primitive p-th root ζ of 1; it is a Galois extension of $\mathbf{Q}$, cyclic of degree $p-1$. The ideal class group A of K is a finite commutative group which measures the failure of the ring of integers of K to be factorial or principal. We are interested in the $\mathbf{F}_p$-space $C = A/A^p$. As A and C are trival for $p = 2$, we assume that the prime p is *odd*.

In the middle of the XIXth century, Kummer made a deep connection between the group C and the sequence of Bernoulli numbers $(B_k)_{k\in\mathbf{N}}$ defined by

$$\frac{T}{e^T - 1} = \sum_{k\in\mathbf{N}} B_k \frac{T^k}{k!},$$

so that $B_0 = 1$, $B_1 = -1/2$. As the function $\frac{T}{e^T-1} - 1 + \frac{T}{2}$ is even (invariant under $T \mapsto -T$), we have $B_k = 0$ for k odd > 1. Here are the numerators N_k and the denominators D_k of the first few Bernoulli numbers $B_k = N_k/D_k$ of even index $k > 0$:

$k = 2$	4	6	8	10	12	14	16	18	20
$N_k = 1$	-1	1	-1	5	-691	7	-3617	43867	-174611
$D_k = 6$	30	42	30	66	2730	6	510	798	330

Kummer's criterion says that *the* $\mathbf{F}_p$*-space* C *is* $\neq 0$ *if and only if* p *divides (the numerator of) one of the rational numbers* $B_2, B_4, \ldots, B_{p-3}$. This is equivalent (see §7) to p dividing the product $B_2B_4 \ldots B_{p-3}$; the sequence of such *irregular* primes is 37 (which divides $B_{32} = 7709321041217/510$), $59|B_{44}$, $67|B_{58}$, $101|B_{68}$, $103|B_{24}, \ldots, 691|B_{12}, \ldots$

The $\mathbf{F}_p$-space C carries an action of the group $\Delta = \mathrm{Gal}(K|\mathbf{Q})$, for which there is a canonical isomorphism $\chi : \Delta \to \mathbf{F}_p^\times$; its inverse sends $a \in \mathbf{F}_p^\times$ to the automorphism $\sigma_a : \zeta \mapsto \zeta^a$ of K. Since every character $\Delta \to \bar{\mathbf{F}}_p^\times$ is of the form χ^i for a unique $i \in \mathbf{Z}/(p-1)\mathbf{Z}$, we have a direct sum decomposition $C = \oplus_{i \in \mathbf{Z}/(p-1)\mathbf{Z}} C(\chi^i)$, where

$$C(\chi^i) = \{c \in C | \sigma(c) = \chi^i(\sigma)c \text{ for every } \sigma \in \Delta\}.$$

In the early 30s, Herbrand refined one of the implications in Kummer's criterion. He showed that, *if* $k \in [2, p-3]$ *is an even integer such that* $C(\chi^{1-k}) \neq 0$, *then* p *divides (the numerator of)* B_k. This is an easy consequence of Stickelberger's theorem [18].

The main result in Ribet's paper is the converse of Herbrand's theorem. In other words, he showed that *if* p *divides* B_k *for some even integer* $k \in [2, p-3]$, *then* $C(\chi^{1-k}) \neq 0$.

(We may allow $k = p-1$ in the statement "$p|B_k \Leftrightarrow C(\chi^{1-k}) \neq 0$" of the Herbrand-Ribet theorem. We shall see in §7 that in fact p always divides the *denominator* of B_{p-1}. We must therefore show that $C(\chi) = 0$. This follows from Kummer theory and the fact that the only degree-p cyclic extension of $\mathbf{Q}$ unramified outside p is the one contained in $\mathbf{Q}(\sqrt[p]{\zeta})$.)

Let us end this introduction by mentioning how the Main Conjecture (for the cyclotomic $\mathbf{Z}_p$-extension of $\mathbf{Q}$), as first proved by Mazur-Wiles, implies a quantitative generalisation of the Herbrand-Ribet theorem; see Ribet's own account [14], also available on his homepage.

Instead of the group $C = A \otimes \mathbf{F}_p$, consider the p-part $D = A \otimes \mathbf{Z}_p$ of the ideal class group A of $K = \mathbf{Q}(\zeta)$. As for C, we have the decomposition $D = \oplus_{i \in \mathbf{Z}/(p-1)\mathbf{Z}} D(\omega^i)$ as a Δ-module, where $\omega : \Delta \to \mathbf{Z}_p^\times$ is obtained from χ by composing with the canonical ("Teichmüller") section $\mathbf{F}_p^\times \to \mathbf{Z}_p^\times$ of the projection $\mathbf{Z}_p^\times \to \mathbf{F}_p^\times$; every character $\Delta \to \bar{\mathbf{Q}}_p^\times$ is of the form ω^i for a unique $i \in \mathbf{Z}/(p-1)\mathbf{Z}$. The Herbrand-Ribet theorem can be restated as "$p|B_k \Leftrightarrow D(\omega^{1-k}) \neq 0$".

The Main Conjecture implies that, $k \in [2, p-3]$ being an even integer, *the group* $D(\omega^{1-k})$ *has the same order as* $\mathbf{Z}_p/L(\omega^{k-1}, 0)\mathbf{Z}_p$, where, for any power ε of ω,

$$L(\varepsilon, 0) = \frac{-1}{p} \sum_{a=1}^{p-1} \varepsilon(\sigma_a) a, \qquad (\sigma_a : \zeta \mapsto \zeta^a).$$

Because of the Kummer congruence $L(\omega^{k-1}, 0) \equiv -B_k/k \pmod p$ [18], we have $p|B_k \Leftrightarrow p|L(\omega^{k-1}, 0)$, so Herbrand-Ribet is indeed a consequence of Mazur-Wiles. (There is as yet no unconditional proof that the groups $D(\omega^{1-k})$ are cyclic.)

Although it is now possible, after the work of Thaine and Kolyvagin, to give a more elementary proof [18] of Herbrand-Ribet, and indeed of Mazur-Wiles, the original approach using Galois representations attached to modular forms was a major source of inspiration for much subsequent work and remains of paramount importance today.

Our aim here is to provide the reader with an overview of Ribet's modular construction (§2–5); the numbering of the statements is the same as in [17]. §1 contains some results of a general nature which are needed in the proof. §6 mentions an alternative approach which can now be carried out. Finally, §7 gives Witt's proof of the von Staudt–Clausen theorem on the denominators of B_k.

1. Prerequisites

In this section we recall, without proof but with references to the literature, a certain number of results which Ribet needs in order to carry out his construction. The reader may go directly to §2–5 and consult this § only if he needs to look up the statement of a result being used in the proof.

Equivariance of the reciprocity map. First, a special case of a certain functorial property of the Artin reciprocity map. Let $M|K|\mathbf{Q}$ be finite extensions such that $M|\mathbf{Q}$ is Galois with group G, such that $K|\mathbf{Q}$ is Galois with group Δ, and such that the Galois group $H \subset G$ of $M|K$ is commutative. Denote by $\mathcal{C}_K$ the idèle class group of K, and by $\psi : \mathcal{C}_K \to H$ the reciprocity map. The group Δ acts on $\mathcal{C}_K$ as well as on H, by conjugation on the latter. We will need the fact that ψ is equivariant for these actions:

Theorem 1.1 ([3]). *For every $\sigma \in G$ (with image $\bar\sigma \in \Delta$) and every $x \in \mathcal{C}_K$, we have $\psi(\bar\sigma x) = \sigma\,\psi(x)\sigma^{-1}$ in H.*

Locally algebraic characters. Let $\bar{\mathbf{Q}}$ be an algebraic closure of $\mathbf{Q}$ and put $G_{\bar{\mathbf{Q}}} = \mathrm{Gal}(\bar{\mathbf{Q}}|\mathbf{Q})$. We refer the reader to [17] for the definition of a *locally algebraic* abelian representation $G_{\bar{\mathbf{Q}}} \to GL_n(K)$ where K is now a finite extension of $\mathbf{Q}_p$. Denote the p-adic cyclotomic character by $\tilde\chi : G_{\bar{\mathbf{Q}}} \to \mathbf{Z}_p^\times \to K^\times$.

Theorem 1.2. *Every locally algebraic character $\varphi : G_{\bar{\mathbf{Q}}} \to K^\times$ is of the form $\varphi = \tilde\chi^n\omega$ for some $n \in \mathbf{Z}$ and some character $\omega : G_{\bar{\mathbf{Q}}} \to K^\times$ of finite order.*

The semisimplification of the reduction of a p-adic representation. The next result says that the reduction of a p-adic representation may depend upon the choice of a stable lattice, but the semisimplification does not. More precisely, let K be a finite extension of $\mathbf{Q}_p$, $\mathfrak{o}$ the ring of integers of K, $\mathfrak{p}$ the unique maximal ideal of $\mathfrak{o}$, and $F = \mathfrak{o}/\mathfrak{p}$ the residue field. Recall that an $\mathfrak{o}$-lattice in a finite-dimensional K-space V is a sub-$\mathfrak{o}$-module generated by some K-basis of V.

Let G be a profinite group — a compact totally disconnected group — and $\rho : G \to GL(V)$ a (continuous) representation. Among the $\mathfrak{o}$-lattices in V, there is at least one which is G-stable, for if $\Lambda \subset V$ is any $\mathfrak{o}$-lattice, the subgroup $H = \rho^{-1}(GL(\Lambda))$ is open in G, hence the index $(G : H)$ is finite, and the $\mathfrak{o}$-lattice $\sum_{\sigma \in G/H}\sigma(\Lambda)$ is G-stable.

The choice of a G-stable $\mathfrak{o}$-lattice $\Lambda \subset V$ leads to a (continuous) representation $\rho_\Lambda : G \to GL(\Lambda) \to GL(\Lambda/\mathfrak{p}\Lambda)$ on a finite-dimensional F-space, called the reduction of ρ along Λ. The theorem of Brauer-Nesbitt says that

Theorem 1.3 ([4]). *The semisimplification $\bar\rho$ of ρ_Λ depends only on ρ, not on the choice of Λ.*

The Galois representation associated to a weight-2 level-p cuspidal eigenform. Next, we need some facts about modular forms; for $N > 0$ we have the congruence subgroups $\Gamma_1(N) \subset \Gamma_0(N) \subset \mathbf{SL}_2(\mathbf{Z})$ whose elements are $\equiv \left(\begin{smallmatrix}1 & \star\\ 0 & 1\end{smallmatrix}\right)$ and $\equiv \left(\begin{smallmatrix}\star & \star\\ 0 & \star\end{smallmatrix}\right)$ (mod N) respectively. For $k > 1$, we denote by $\mathfrak{S}_k(N)$ the space of cuspidal modular forms of weight k for $\Gamma_1(N)$, and by $\mathfrak{S}_k(N, \varepsilon)$ the subspace on which

$\Gamma_0(N)/\Gamma_1(N) = (\mathbf{Z}/N\mathbf{Z})^\times$ acts via a given character $\varepsilon : (\mathbf{Z}/N\mathbf{Z})^\times \to \mathbf{C}^\times$, so that

$$\mathfrak{S}_k(N) = \bigoplus_{\varepsilon:(\mathbf{Z}/N\mathbf{Z})^\times \to \mathbf{C}^\times} \mathfrak{S}_k(N, \varepsilon).$$

When $\varepsilon = 1$ is trivial, $\mathfrak{S}_k(N, 1)$ is the space of weight-k cuspidal modular forms for $\Gamma_0(N)$.

Let $f \in \mathfrak{S}_2(p)$ be a normalised weight-2 cuspidal eigenform of level p, and write the q-expansion (with $q = e^{2i\pi\tau}$ and $\tau \in \mathfrak{H}$) of f as $f = \sum_{n>0} a_n q^n$.

Theorem 1.4 ([8]). *The subfield $K_f = \mathbf{Q}(a_1, a_2, \dots)$ of $\mathbf{C}$ is of finite degree over $\mathbf{Q}$. Moreover, if $\varepsilon : (\mathbf{Z}/p\mathbf{Z})^\times \to \mathbf{C}^\times$ is the character such that $f \in \mathfrak{S}_2(p, \varepsilon)$, then the field $\mathbf{Q}(\varepsilon)$ generated by the values of ε is contained in the number field K_f.*

Recall the construction of the (2-dimensional) representations $\rho_{f,\mathfrak{p}} : G_{\bar{\mathbf{Q}}} \to GL(V_{f,\mathfrak{p}})$ attached to a normalised eigenform $f \in \mathfrak{S}_2(p, \varepsilon)$. Here $\mathfrak{p}$ is any prime of K_f and $V_{f,\mathfrak{p}}$ is a certain vector space — described below — over the completion $K_{f,\mathfrak{p}}$ of K_f at the prime $\mathfrak{p}$. Shimura attaches to f an abelian variety A_f over $\mathbf{Q}$ (uniquely determined up to $\mathbf{Q}$-isogeny) with the following properties.

The abelian variety A_f has dimension $[K_f : \mathbf{Q}]$; it is a quotient of the jacobian $J_1(p) = \mathrm{Jac}X_1(p)$ of the modular curve $X_1(p)$; and there is an embedding of $\mathbf{Q}$-algebras $K_f \to \mathrm{End}_{\mathbf{Q}}(A_f) \otimes \mathbf{Q}$.

Let $T_p(A_f) = \lim_{n\in\mathbf{N}^*} {}_{p^n}A_f(\bar{\mathbf{Q}})$ be the p-adic Tate module of A_f and $V_p(A_f) = T_p(A_f) \otimes_{\mathbf{Z}_p} \mathbf{Q}_p$, which is thus a module over $K_f \otimes \mathbf{Q}_p$.

Theorem 1.5 ([8]). *The $(K_f \otimes \mathbf{Q}_p)$-module $V_p(A_f)$ is free of rank* 2.

It follows that $V_{f,\mathfrak{p}} = V_p(A_f) \otimes_{K_f \otimes \mathbf{Q}_p} K_{f,\mathfrak{p}}$ is a 2-dimensional vector space over $K_{f,\mathfrak{p}}$ with a continuous linear action of $G_{\bar{\mathbf{Q}}}$. Denote this degree-2 representation by

$$\rho_{f,\mathfrak{p}} : G_{\bar{\mathbf{Q}}} \to GL(V_{f,\mathfrak{p}}).$$

The abelian variety A_f has good reduction at every prime $l \neq p$ of $\mathbf{Q}$. Further, the Eichler-Shimura relations give

Theorem 1.6 ([8]). *The representation $\rho_{f,\mathfrak{p}}$ is unramified at every prime $l \neq p$ and the trace (resp. determinant) of a Frobenius element* Frob_l *at l is a_l (resp. $l\varepsilon(l)$) in $K_{f,\mathfrak{p}}$ (resp. $K_{f,\mathfrak{p}}^\times$).*

The final bit of information about A_f that we need is the following theorem of Deligne-Rapoport.

Theorem 1.7 ([6]). *Suppose that ε is not trivial. The abelian variety A_f acquires good reduction at the unique prime dividing p in the fixed field $\mathbf{Q}(\zeta)^+ \subset \mathbf{Q}(\zeta)$ (where $\zeta^p = 1$, $\zeta \neq 1$) of the involution $\zeta \mapsto \zeta^{-1}$.*

Finite flat group schemes. Let $E|\mathbf{Q}_p$ be a finite extension, with ring of integers $\mathfrak{o}_E$. Ribet makes crucial use of the following result of Raynaud.

Theorem 1.8 ([12]). *Suppose that the ramification index of $E|\mathbf{Q}_p$ is $< p - 1$. Let G be a finite flat commutative group scheme over E, killed by a power of p. Then there is at most one finite flat extension of G to $\mathfrak{o}_E$.*

2. The main theorem

Adopt the same notation as in the introduction. In particular, p is an odd prime and $K = \mathbf{Q}(\zeta)$, with $\zeta^p = 1$, $\zeta \neq 1$. We reduce the main theorem (2.1) to the existence of a suitable finite extension E (2.2) of $\mathbf{Q}$, which in turn follows from the construction of a suitable representation r (2.3) of $\mathrm{Gal}(\bar{\mathbf{Q}}|\mathbf{Q})$. The existence of r is established in §5, as a suitable reduction (§3) of the representation associated to a certain $f \in \mathfrak{S}_2(p, \varepsilon)$ for a suitable ε (§4).

Theorem 2.1. *Let* $k \in [2, p-3]$ *be an even integer. If* $p|B_k$, *then* $C(\chi^{1-k}) \neq 0$.

By the functorial property (1.1) of the Artin symbol, we are reduced to constructing a suitable everywhere-unramified extension $E|K$.

Theorem 2.2. *Let* $k \in [2, p-3]$ *be an even integer, and suppose that* $p|B_k$. *There exists a Galois extension* $E|\mathbf{Q}$ *containing* K *such that*

(a) *The extension* $E|K$ *is everywhere unramified.*
(b) *The group* $H = \mathrm{Gal}(E|K)$ *is a p-elementary commutative group* ($\neq 1$)*; equivalently,* H *is an* $\mathbf{F}_p$*-space of dimension* > 0.
(c) *For every* σ *in* $G = \mathrm{Gal}(E|\mathbf{Q})$ *(with image* $\bar{\sigma}$ *in* Δ*) and every* τ *in* H,

$$\sigma\tau\sigma^{-1} = \chi(\bar{\sigma})^{1-k} \cdot \tau$$

Proof that (2.2) $\Rightarrow$ (2.1). Let M be the maximal p-elementary abelian extension of K which is unramified at every place of K; by maximality, $M|\mathbf{Q}$ is Galois. The Artin map $\psi : C_K \to \mathrm{Gal}(M|K)$, where C_K is the idèle class group of K, is a homomorphism inducing an isomorphism $C \to \mathrm{Gal}(M|K)$. Moreover, for every $\sigma \in \mathrm{Gal}(M|\mathbf{Q})$ (with image $\bar{\sigma} \in \Delta$) and every $x \in C_K$, we have $\psi(\bar{\sigma}x) = \sigma\psi(x)\sigma^{-1}$ (0.1).

In other words, the isomorphism (of $\mathbf{F}_p$-spaces) $C \to \mathrm{Gal}(M|K)$ induced by ψ is Δ-equivariant, and hence induces, for every $i \in \mathbf{Z}/(p-1)\mathbf{Z}$, an isomorphism $C(\chi^i) \to \mathrm{Gal}(M|K)(\chi^i)$. If (2.2) holds, the latter group is $\neq 0$ for $i = 1-k$, because it has the quotient H. It follows that if $p|B_k$, then $C(\chi^{1-k}) \neq 0$, so (2.1) holds (if (2.2) holds).

The next theorem shows how such an E can be constructed from a certain kind of representation of $G_{\bar{\mathbf{Q}}}$, where $\bar{\mathbf{Q}}$ is an algebraic closure of $\mathbf{Q}$ containing K. Choose a decomposition group $D_p \subset G_{\bar{\mathbf{Q}}}$ at the prime p, and let χ also stand for the composite $G_{\bar{\mathbf{Q}}} \to \Delta \to \mathbf{F}_p^\times$.

Theorem 2.3. *Let* $k \in [2, p-3]$ *be even, and suppose that* $p|B_k$. *There exists a finite extension* $F|\mathbf{F}_p$, *and a (continuous) representation* $r : G_{\bar{\mathbf{Q}}} \to GL_2(F)$, *such that*

(i) r *is unramified at every prime* $l \neq p$.
(ii) r *is an extension of* χ^{k-1} *by the trivial* 1*-dimensional representation:*

$$r \sim \begin{pmatrix} 1 & \gamma \\ 0 & \chi^{k-1} \end{pmatrix}$$

for some map $\gamma : G_{\bar{\mathbf{Q}}} \to F$.

(iii) *The order of* $\mathrm{Im}(r)$ *is divisible by* p.
(iv) *The order of* $\mathrm{Im}(r|_{D_p})$ *is not divisible by* p.

Proof that (2.3) $\Rightarrow$ (2.2). Let r be a representation as in (2.3); we have $\det(r) = \chi^{k-1}$, and there is a tower of fields $E'|K'|\mathbf{Q}$, with E' the field fixed by $\mathrm{Ker}(r)$ and K' the field fixed by $\mathrm{Ker}(\chi^{k-1})$; we have $K' \subset K$, because $\mathrm{Ker}(\chi) \subset \mathrm{Ker}(\chi^{k-1})$. Put $H' = \mathrm{Gal}(E'|K')$, so that the sequence $1 \to H' \to \mathrm{Im}(r) \to \mathrm{Im}(\chi^{k-1}) \to 1$ is exact. Notice that H' is a p-elementary commutative group, for γ gives rise to an injective homomorphism $H' \to F$, and its order is > 1, for the order of $\mathrm{Im}(r)$ is divisible by p, by (2.3)(iii).

Next, we remark that $E'|K'$ is everywhere unramified. Indeed, for every prime $l \neq p$, the extension $E'|\mathbf{Q}$ is unramified at l by (2.3)(i), and hence so is $E'|K'$ at every prime $\mathfrak{l}$ of K' prime to p. As for the unique prime $\mathfrak{p}$ of K' dividing p, the inertia subgroup at $\mathfrak{p}$ has order prime to p by (2.3)(iv), so $E'|K'$ is at worst tamely ramified, and hence unramified, because $E'|K'$ is a p-extension.

Taking $E = E'K$, it is now clear that (2.2)(a) and (2.2)(b) hold for $E|K|\mathbf{Q}$, because the analogous statements hold for $E'|K'|\mathbf{Q}$, as we have just seen. It remains to show that (2.2)(c) holds as well. This follows from the fact that for every $\sigma, \tau \in G_{\bar{\mathbf{Q}}}$, we have

$$\begin{pmatrix}1 & \gamma(\sigma)\\ 0 & \chi^{k-1}(\sigma)\end{pmatrix}\begin{pmatrix}1 & \gamma(\tau)\\ 0 & 1\end{pmatrix}\begin{pmatrix}1 & \gamma(\sigma)\\ 0 & \chi^{k-1}(\sigma)\end{pmatrix}^{-1} = \begin{pmatrix}1 & \chi^{1-k}(\sigma)\cdot\gamma(\tau)\\ 0 & 1\end{pmatrix},$$

because of the identity

$$\begin{pmatrix}a & b\\ 0 & d\end{pmatrix}\begin{pmatrix}1 & x\\ 0 & 1\end{pmatrix}\begin{pmatrix}a & b\\ 0 & d\end{pmatrix}^{-1} = \begin{pmatrix}1 & ad^{-1}\cdot x\\ 0 & 1\end{pmatrix}.$$

Thus the implications (2.3) $\Rightarrow$ (2.2) $\Rightarrow$ (2.1) are proved. It remains to find an r as in (2.3) to complete the proof of the main theorem (2.1). We shall take r to be a suitable reduction (§3) of the representation $\rho_{f,\mathfrak{p}}$ (§5) attached to a certain weight-2 level-p cuspidal eigenform f (§4).

3. Reductions of p-adic representations

Let $K|\mathbf{Q}_p$ be a finite extension, $\mathfrak{o}$ its ring of integers, $\mathfrak{p}$ the maximal ideal of $\mathfrak{o}$, and F the residue field. Let V be a 2-dimensional K-space with a continuous linear action ρ of a profinite group G, and suppose that $\bar{\rho}$ (1.3) is reducible. The two characters $\varphi_1, \varphi_2 : G \to F^\times$ such that $\bar{\rho} \sim \varphi_1 \oplus \varphi_2$ are uniquely determined by ρ up to order. Also, for every G-stable $\mathfrak{o}$-lattice $\Lambda \subset V$, the reduction ρ_Λ is reducible, and its semisimplification is isomorphic to $\varphi_1 \oplus \varphi_2$ (1.3). For a given Λ, ρ_Λ is isomorphic to one of

$$\begin{pmatrix}\varphi_1 & \star\\ 0 & \varphi_2\end{pmatrix},\quad \begin{pmatrix}\varphi_1 & 0\\ \star & \varphi_2\end{pmatrix},$$

and ρ_Λ is semisimple if $\rho_\Lambda(G)$ has order prime to p. Ribet's lemma says that each of these two types can be realised for some Λ, and, if ρ is simple (but $\bar{\rho}$ is not), then there is a ρ_Λ which is not semisimple and has any of the displayed types prescribed in advance.

Theorem 3.1 ("Ribet's lemma"). *Suppose that the degree-2 representation V of G is simple but $\bar{\rho}$ is not simple. Then, for any ordering φ_1, φ_2 of the two characters of which $\bar{\rho}$ is the direct sum, there is a G-stable $\mathfrak{o}$-lattice $\Lambda \subset V$ such that $\rho_\Lambda \sim \begin{pmatrix} \varphi_1 & \star \\ 0 & \varphi_2 \end{pmatrix}$, as opposed to $\begin{pmatrix} \varphi_1 & 0 \\ \star & \varphi_2 \end{pmatrix}$, and such that ρ_Λ is not semisimple, i.e., $\rho_\Lambda \not\sim \varphi_1 \oplus \varphi_2$.*

Choosing a G-stable $\mathfrak{o}$-lattice $\Lambda \subset V$ and fixing an $\mathfrak{o}$-basis of Λ allows one to view ρ as a representation $G \to GL_2(\mathfrak{o})$; the reduction of ρ along Λ is then the composite $\rho_\Lambda : G \to GL_2(\mathfrak{o}) \to GL_2(F)$; we call it the reduction of $\rho(G)$. For every $M \in GL_2(K)$ such that $M\rho(G)M^{-1} \subset GL_2(\mathfrak{o})$, we get another G-stable $\mathfrak{o}$-lattice in V together with an $\mathfrak{o}$-basis, giving another reduction $G \to M\rho(G)M^{-1} \to GL_2(\mathfrak{o}) \to GL_2(F)$ of ρ. We will show that at least one of these reductions is of the desired type $\begin{pmatrix} \varphi_1 & \star \\ 0 & \varphi_2 \end{pmatrix}$, and that, if all reductions of this type are semisimple, then ρ cannot be simple, which it is by hypothesis.

Let π be an $\mathfrak{o}$-basis of $\mathfrak{p}$ (a uniformiser of K), and put $P = \begin{pmatrix} 1 & 0 \\ 0 & \pi \end{pmatrix}$; we have

$$P\begin{pmatrix} a & \pi b \\ c & d \end{pmatrix}P^{-1} = \begin{pmatrix} a & b \\ \pi c & d \end{pmatrix}.$$

This implies that there is at least one G-stable $\mathfrak{o}$-lattice in V along which the reduction of ρ is of the desired type $\begin{pmatrix} \varphi_1 & \star \\ 0 & \varphi_2 \end{pmatrix}$, for if the reduction of $\rho(G)$ is of the other type $\begin{pmatrix} \varphi_1 & 0 \\ \star & \varphi_2 \end{pmatrix}$, then the reduction of $P\rho(G)P^{-1}$ is of the desired type, as the above formula shows.

Suppose therefore that the reduction of $\rho(G)$ is of the desired type $\begin{pmatrix} \varphi_1 & \star \\ 0 & \varphi_2 \end{pmatrix}$; it remains to show that, among the reductions of ρ of this type, there is at least one which is not semisimple. Suppose, if possible, that every reduction of the type $\begin{pmatrix} \varphi_1 & \star \\ 0 & \varphi_2 \end{pmatrix}$ is semisimple. We shall construct a sequence of matrices $(M_i)_{i\in\mathbf{N}}$ in $GL_2(\mathfrak{o})$, having a limit $M \in GL_2(\mathfrak{o})$, such that $M\rho(G)M^{-1} \subset \begin{pmatrix} \mathfrak{o}^\times & 0 \\ \mathfrak{p} & \mathfrak{o}^\times \end{pmatrix}$, showing that ρ is not simple — a contradiction.

Suppose that, for some $i \in \mathbf{N}$, a matrix $M_i = \begin{pmatrix} 1 & t_i \\ 0 & 1 \end{pmatrix}$ in $GL_2(\mathfrak{o})$ has been found such that

$$M_i\rho(G)M_i^{-1} \subset \begin{pmatrix} \mathfrak{o}^\times & \mathfrak{p}^i \\ \mathfrak{p} & \mathfrak{o}^\times \end{pmatrix};$$

for $i = 0$, take $M_0 = \begin{pmatrix} 1 & 0 \\ 0 & 1 \end{pmatrix}$. The above inclusion can be rewritten as $P^i M_i\rho(G)M_i^{-1} P^{-i} \subset \begin{pmatrix} \mathfrak{o}^\times & \mathfrak{o} \\ \mathfrak{p}^{i+1} & \mathfrak{o}^\times \end{pmatrix}$, whose reduction is of the desired type $\begin{pmatrix} \varphi_1 & \star \\ 0 & \varphi_2 \end{pmatrix}$, because the reduction of $\rho(G)$ is of this type. As all reductions of this type are semisimple by hypothesis, there exists a matrix $U_i = \begin{pmatrix} 1 & u_i \\ 0 & 1 \end{pmatrix}$ in $GL_2(\mathfrak{o})$ which diagonalises the reduction in the sense that

$$U_i P^i M_i \rho(G) M_i^{-1} P^{-i} U_i^{-1} \subset \begin{pmatrix} \mathfrak{o}^\times & \mathfrak{p} \\ \mathfrak{p} & \mathfrak{o}^\times \end{pmatrix}.$$

Denoting the entries of a matrix $() \in GL_2(\mathfrak{o})$ by $(a_{11}(), a_{12}(); a_{21}(), a_{22}())$, one has $a_{21}(U_i M U_i^{-1}) = a_{21}(M)$. This means that we actually have

$$U_i P^i M_i \rho(G) M_i^{-1} P^{-i} U_i^{-1} \subset \begin{pmatrix} \mathfrak{o}^\times & \mathfrak{p} \\ \mathfrak{p}^{i+1} & \mathfrak{o}^\times \end{pmatrix},$$

for the a_{21} of the matrix being conjugated by U_i was in $\mathfrak{p}^{i+1}$. Conjugating by P^{-i} and taking $M_{i+1} = P^{-i} U_i P^i M_i = \begin{pmatrix} 1 & t_i + u_i \pi^i \\ 0 & 1 \end{pmatrix}$, the above inclusion implies

$$M_{i+1} \rho(G) M_{i+1}^{-1} \subset \begin{pmatrix} \mathfrak{o}^\times & \mathfrak{p}^{i+1} \\ \mathfrak{p} & \mathfrak{o}^\times \end{pmatrix},$$

completing the construction of the sequence $(M_i)_{i \in \mathbf{N}}$ by induction. It is clear that this sequence converges in $GL_2(\mathfrak{o})$, and that the limit M satisfies $M\rho(G)M^{-1} \subset \begin{pmatrix} \mathfrak{o}^\times & 0 \\ \mathfrak{p} & \mathfrak{o}^\times \end{pmatrix}$, which contradicts the simplicity of ρ and thereby proves that ρ has at least one reduction which has the desired type but which is not semisimple.

For a generalisation of (3.1), see [1].

4. A weight-2 level-p cuspidal eigenform

Let p be an odd prime, and let $\omega : \mathbf{F}_p^\times \to \mathbf{Z}_p^\times$ be the unique character (the Teichmüller lift) such that $\omega(d) \equiv d \pmod{p}$ for every $d \in \mathbf{F}_p^\times$. The image of ω is the subgroup ${}_{p-1}\mu$ or roots of 1 of order dividing $p-1$. If a number field K contains these roots of 1, and if we fix a prime $\mathfrak{p}|p$ of K, then we may think of ω as taking values in $K^\times$.

Let $k \in [2, p-3]$ be an even integer such that $p|B_k$, and put $\varepsilon = \omega^{k-2}$. See Anupam Saikia's contribution [15] to this volume for the proof of the following crucial result.

Theorem 4.1. *Suppose that $p|B_k$. There exist a normalised cuspidal eigenform $f \in \mathfrak{S}_2(p, \varepsilon)$, $f = \sum_{n>0} a_n q^n$, of weight 2, level p, and character ε, and a prime $\mathfrak{p}|p$ of the number field K_f (0.4) generated by the q-coefficients of f such that for every prime $l \neq p$, the number a_l is $\mathfrak{p}$-integral and*

$$a_l \equiv 1 + l^{k-1} \equiv 1 + \varepsilon(l) l \pmod{\mathfrak{p}}.$$

Note that K_f is a number field (1.4) and that ε may be thought of as taking values in $K_f^\times$, for K_f contains $\mathbf{Q}(\varepsilon)$, the number field generated by the values of ε (1.4).

Now let $\rho_{f,\mathfrak{p}}$ be the representation attached to f at $\mathfrak{p}$ (§1); we show (§5) that it has a reduction r having the properties (i)–(iv) of (2.3).

5. The Galois representation

Denoting the p-adic cyclotomic character by $\tilde{\chi} : G_{\bar{\mathbf{Q}}} \to \mathbf{Z}_p^\times$ (so that $\chi : \Delta \to \mathbf{F}_p^\times$ is the reduction of $\tilde{\chi}$), and thinking of it as taking values in $K_{f,\mathfrak{p}}^\times$, the equality $\det(\rho_{f,\mathfrak{p}}(\mathrm{Frob}_l)) = l\varepsilon(l)$ (1.6) can be rewritten as $\det \rho_{f,\mathfrak{p}} = \tilde{\chi} \cdot \varepsilon$.

Theorem 5.1. *The representation* $\rho_{f,\mathfrak{p}}$ *is (absolutely) simple.*

Suppose that it is not; its semisimplification is then $\rho_1 \oplus \rho_2$, for some characters $\rho_i : G_{\bar{\mathbf{Q}}} \to K_{f,\mathfrak{p}}^\times$. It follows (1.2) that each ρ_i can be written as χ^{n_i} (for some $n_i \in \mathbf{Z}$) on an open subgroup of the inertia subgroup at p, and therefore $\rho_i = \varepsilon_i \chi^{n_i}$ for some characters of finite order $\varepsilon_i : G_{\bar{\mathbf{Q}}} \to K_{f,\mathfrak{p}}^\times$ unramified away from p.

Thinking of the ε_i as characters of $(\mathbf{Z}/m_i\mathbf{Z})^\times$ for some m_i, we have, for every prime $l \neq p$, the relations

$$\varepsilon_1(l)l^{n_1} + \varepsilon_2(l)l^{n_2} = a_l, \quad \varepsilon_1(l)\varepsilon_2(l)l^{n_1+n_2} = l\varepsilon(l).$$

From the second equation we get $n_1 + n_2 = 1$, and hence one of the n_i is ≥ 1 and the other ≤ 0. The first equation then gives $|a_l| \geq l - 1$ which, for $l \geq 7$, contradicts the "Riemann hypothesis" $|a_l| \leq 2\sqrt{l}$, as proved by Eichler-Shimura.

Denote the ring of integer of $K_{f,\mathfrak{p}}$ by $\mathfrak{o}_{f,\mathfrak{p}}$.

Theorem 5.2. *There exists a* $G_{\bar{\mathbf{Q}}}$*-stable* $\mathfrak{o}_{f,\mathfrak{p}}$*-lattice* $\Lambda \subset V_{f,\mathfrak{p}}$ *such that*

$$\rho_{f,\mathfrak{p},\Lambda} \sim \begin{pmatrix} 1 & \star \\ 0 & \chi^{k-1} \end{pmatrix}, \quad \rho_{f,\mathfrak{p},\Lambda} \not\sim \begin{pmatrix} 1 & 0 \\ 0 & \chi^{k-1} \end{pmatrix}.$$

We know that $\rho_{f,\mathfrak{p}}$ is irreducible (5.1). The proposition would follow from (3.1) once we show that $\bar{\rho}_{f,\mathfrak{p}} \sim 1 \oplus \chi^{k-1}$, where $\bar{\rho}_{f,\mathfrak{p}}$ is the semisimplification of $\rho_{f,\mathfrak{p},\Lambda}$ (1.3).

Indeed, for every prime $l \neq p$, we know (1.6) that $\rho_{f,\mathfrak{p}}$, and hence the reduction $\bar{\rho}_{f,\mathfrak{p}}$, is unramified at l. Moreover (1.6), the trace and the determinant of $\mathrm{Frob}_l \in \mathrm{Im}\rho_{f,\mathfrak{p}}$ are a_l and $l\varepsilon(l)$ respectively. By (4.7) and the definition of ε, these numbers are congruent respectively to $1 + l^{k-1}$ and l^{k-1} modulo $\mathfrak{p}$. By the Chebotarev density theorem, the trace and the determinant of $\bar{\rho}_{f,\mathfrak{p}}$ are $1 + \chi^{k-1}$ and χ^{k-1} respectively, which are the same as the trace and the determinant of the (semisimple) representation $1 \oplus \chi^{k-1}$. But two degree-2 semisimple representations having the same trace and the same determinant are the same (1.3), hence $\bar{\rho}_{f,\mathfrak{p}} \sim 1 \oplus \chi^{k-1}$, which was to be proved.

Now fix such a lattice Λ and let $r = \rho_{f,\mathfrak{p},\Lambda}$ be the reduction of $\rho_{f,\mathfrak{p}}$ along Λ. We have seen that the representation r satisfies the conditions (i)–(iii) of (2.3); it remains to check that it satisfies (iv) — the restriction $r|_{D_p}$ of r to a decomposition group D_p at p is semisimple.

Denote by F the residue field of $K_{f,\mathfrak{p}}$ and by M the (2-dimensional) F-space which carries the representation r. Let L be the completion of $\mathbf{Q}(\zeta)^+$ at the unique place above p, and $D = \mathrm{Gal}(\bar{L}|L)$ the decomposition group at the said place. It is sufficient to show that $r|_D$ (or the D-module M) is semisimple, for the index $(D_p : D)$ is prime to p, where we have identified D_p with $\mathrm{Gal}(\bar{L}|\mathbf{Q}_p)$.

Theorem 5.3. *The* D*-module* M *comes from a finite flat* p*-elementary commutative group scheme over the ring of integers* $\mathfrak{o}_L$ *of* L.

There is an abelian $\mathbf{Q}$-variety A which is $\mathbf{Q}$-isogenous to A_f and such that $\mathfrak{o}_f \subset \mathrm{End}(A)$, because $K_f \subset \mathrm{End}_{\mathbf{Q}}(A_f) \otimes \mathbf{Q}$. The D-module M is then the "kernel" ${}_{\mathfrak{p}}A$ of $\mathfrak{p}$, and hence contained in the D-module ${}_pA$, the kernel of p. As A has good reduction over L (1.7), the D-module ${}_pA$ comes from a finite flat p-elementary commutative group scheme over $\mathfrak{o}_L$, namely the kernel ${}_p\mathcal{A}$ of p, where $\mathcal{A}$ — the Néron model of A — is the abelian $\mathfrak{o}_L$-scheme whose generic fibre is A. It follows that the D-module $M = {}_{\mathfrak{p}}A$ comes from the the schematic closure $\mathcal{M}$ of M in ${}_p\mathcal{A}$ (1.8).

Theorem 5.4. *The D-module M is semisimple.*

This is the same as saying that the order of $r(D)$ is prime to p, and will follow from the existence of two distinct F-lines (X and M^0) in M which are D-stable.

We know that r, and hence $r|_D$, is isomorphic to $\begin{pmatrix} 1 & \star \\ 0 & \chi^{k-1} \end{pmatrix}$. Let $X \subset M$ be an F-line on which D acts trivially, so that D acts via χ^{k-1} on the F-line $Y = M/X$; in particular, Y is ramified.

Let $\mathcal{X}$ be the schematic closure of X in $\mathcal{M}$, so that the D-module coming from $\mathcal{X}$ is X. As the absolute ramification index $(p-1)/2$ of L is $< p-1$, the group scheme $\mathcal{X}$ is constant (1.8). It follows that $\mathcal{M}$ cannot be connected, for it has the étale subgroup scheme $\mathcal{X}$ (of order > 1).

Now, the group scheme $\mathcal{M}$ is an F-space scheme. Let $M^0 \subset M$ be the sub-$F[D]$-module coming from the largest connected subgroup scheme of $\mathcal{M}$, so that the $F[D]$-module M/M^0 comes from the largest étale quotient of $\mathcal{M}$ and hence M/M^0 is unramified.

We have $M^0 \neq M$, because $\mathcal{M}$ is not connected, as we have seen. We have $M^0 \neq 0$ because M/M^0 is unramified whereas M is not (for it has the quotient Y which is ramified). For the same reason, $M^0 \neq X$, because M/M^0 is unramified whereas $Y = M/X$ is ramified. Thus X and M^0 are two distinct D-stable F-lines in M, and hence the D-module M is semisimple, which was to be shown.

Summary of the proof of the main theorem (2.1). The first step (4.7) is to construct, if $p|B_k$, a normalised cuspidal eigenform $f \in \mathfrak{S}_2(p)$ such that for every prime $l \neq p$, the q-coefficient $a_l(f)$ is $\equiv 1 + l^{k-1}$ modulo a certain prime $\mathfrak{p}|p$ of K_f. A certain reduction r (3.1) of the associated $\mathfrak{p}$-adic representation $\rho_{f,\mathfrak{p}}$ gives an everywhere-unramified p-elementary abelian extension $E|\mathbf{Q}(\zeta)$ which is Galois over $\mathbf{Q}$ and such that $\Delta = \mathrm{Gal}(\mathbf{Q}(\zeta)|\mathbf{Q})$ operates on $H = \mathrm{Gal}(E|\mathbf{Q}(\zeta))$ via the character $\chi^{1-k} : \Delta \to \mathbf{F}_p^\times$. Hence $C(\chi^{1-k}) \neq 0$.

6. A weight-*k* level-1 cuspidal eigenform

Ribet was aware at the time of writing [13] of Serre's suggestion [16] that instead of the weight-2 level-p cuspidal eigenform $f \in \mathfrak{S}_2(p, \varepsilon)$ of (4.7), one could work with a suitable weight-k level-1 cuspidal eigenform $g \in \mathfrak{S}_k(1)$. The construction of g is more conceptual than that of f. He could show that for some prime $\mathfrak{q}|p$ of the number field K_g, the associated representation $\rho_{g,\mathfrak{q}}$ has a reduction r satisfying properties (i)–(iii) of (2.3).

However, the techniques of the day were insufficient to prove that there is a reduction satisfying (2.3)(iv) as well. But, as C. Khare [11] points out, it has now become possible to show that $\rho_{g,\mathfrak{q}}$ has a reduction r satisfying all four requirements of (2.3), thanks to the work of Faltings-Jordan [9] and Fontaine-Laffaille [10]. We give the broad outline.

Theorem 6.1. *Suppose that* $p|B_k$ *for some even* $k \in [2, p-3]$. *There exists a normalised weight-k level-1 cuspidal eigenform* $g = \sum_{n>0} a_n(g)q^n$ *in* $\mathfrak{S}_k(1)$ *and a prime* $\mathfrak{q}|p$ *of the number field* K_g *generated by the q-coefficients of* g *such that for every prime* $l \neq p$, *the number* $a_l(g)$ *is* $\mathfrak{q}$*-integral and*

$$a_l(g) \equiv 1 + l^{k-1} \pmod{\mathfrak{q}}.$$

Recall that the Eisenstein series $G_j = (-B_j/2j) + \sum_{n>0} \sigma_{j-1}(n)q^n$, where $\sigma_i(n) = \sum_{d|n} d^i$, is in the space $\mathfrak{M}_j(1)$ of modular forms of weight j (and level 1) for every even integer $j > 2$; our k is > 2, for the numerator of B_2 is 1. Notice that $a_l(G_k) = 1 + l^{k-1}$ for every prime l. Also recall that there is a unique normalised cuspidal eigenform $\Delta \in \mathfrak{S}_{12}(1)$, with q-expansion

$$q \prod_{n>0} (1-q^n)^{24} = \sum_{n>0} \tau(n)q^n.$$

The graded $\mathbf{C}$-algebra of level-1 modular forms $\mathfrak{M}(1)$ is the polynomial ring $\mathbf{C}[G_4, G_6]$, and the graded ideal $\mathfrak{S}(1) \subset \mathfrak{M}(1)$ of level-1 cuspidal forms is principal, generated by Δ [8].

Following a suggestion of K. Joshi, we therefore look for a g of the type $g = g_1\Delta + \cdots + g_n\Delta^n$ for a suitable $n > 0$, where $g_i = a_i G_4^{c_i} G_6^{d_i}$ ($a_i \in \mathbf{Z}$, $c_i, d_i \in \mathbf{N}$, $4c_i + 6d_i + 12i = k$). Notice that every even integer > 2 can be written as $4c + 6d$ for some $c, d \in \mathbf{N}$.

The existence of a $g \in \mathfrak{S}_k(1)_{\mathbf{Z}}$ such that $g \equiv G_k \pmod{p}$ now follows from the fact that the constant terms $-B_4/8$ and $-B_6/12$ of G_4, G_6 have numerators 1, -1 respectively, and that if $f_1, f_2 \in \mathfrak{M}_k(1)$ have the same first few (roughly $k/12$) q-coefficients, then $f_1 = f_2$, because the space $\mathfrak{M}_k(1)$ is finite-dimensional. For example, $g = \Delta$ would do if $p = 691$ and $k = 12$.

But we need a normalised *eigenform*, not just a cuspidal form. This is furnished by the following lemma of Deligne-Serre, which implies the existence of a number field L', a prime $\mathfrak{q}|p$ of L', and a normalised eigenform $g \in \mathfrak{S}_k(1)_{L'}$ with $\mathfrak{q}$-integral q-coefficients, such that the congruence $a_l(g) \equiv 1 + l^{k-1} \pmod{\mathfrak{q}}$ holds for every prime $l \neq p$.

Theorem 6.2 (Deligne-Serre [7]). *Let A be a discrete valuation ring,* $\mathfrak{m}$ *its maximal ideal, and L its field of fractions. Let M be a free A-module of finite rank, and* $\mathfrak{h}$ *a set of commuting A-endomorphisms of M. Let* $f \in M/\mathfrak{m}M$ *be an eigenvector for every* $T \in \mathfrak{h}$, *of eigenvalue* $a_T \in A/\mathfrak{m}$. *There exists a discrete valuation ring* A' *containing A, whose field of fractions is a finite extension of L and whose maximal ideal* $\mathfrak{m}'$ *restricts to* $\mathfrak{m}$ ($\mathfrak{m}' \cap A = \mathfrak{m}$), *and a common eigenvector* $f' \in M \otimes_A A'$ *for* $\mathfrak{h}$ *such that the eigenvalue* $a'_T \in A'$ *of* f' *is* $\equiv a_T \pmod{\mathfrak{m}'}$ *for every* $T \in \mathfrak{h}$.

See Anupam Saikia's contribution to this volume [15] for a proof.

To complete the proof of (6.1), apply this lemma with $A = \mathbf{Z}_{(p)}$, $M = \mathfrak{S}_k(1)_{\mathbf{Z}_{(p)}}$, $\mathfrak{h}$ the set of Hecke operators T_l for primes l, and $f = \bar{g}$, the reduction of g modulo p, which is an eigenform in $M/pM = \mathfrak{S}_k(1)_{\mathbf{F}_p}$, the eigenvalue for T_l being $1 + \bar{l}^{k-1} \in \mathbf{F}_p$ for every $l \neq p$, because $g \equiv G_k$.

Recall that to g and $\mathfrak{q}$ as in (6.1) Deligne [5] attaches a degree-2 representation $\rho_{g,\mathfrak{q}}$ of $G_{\bar{\mathbf{Q}}}$ over the completion $K_{g,\mathfrak{q}}$ of $K_g = \mathbf{Q}(a_n(g))_{n>0}$ at $\mathfrak{q}$; it is unramified at every prime $l \neq p$; the trace of $\rho_{g,\mathfrak{q}}(\mathrm{Frob}_l)$ is $a_l(g)$, the determinant l^{k-1}.

Theorem 6.3. *The representation $\rho_{g,\mathfrak{q}}$ is (absolutely) simple.*

The proof is the same as for the representation $\rho_{f,\mathfrak{p}}$ in (5.1), using the Ramanujan-Peterson estimate "$|a_l(g)| \le 2\sqrt{l^{k-1}}$ for every prime $l \neq p$", as proved by Deligne [5], instead of the earlier "$|a_l(f)| \le 2\sqrt{l}$".

Theorem 6.4. *The representation $\rho_{g,\mathfrak{q}}$ has a reduction r such that*

$$r \sim \begin{pmatrix} 1 & \star \\ 0 & \chi^{k-1} \end{pmatrix}, \quad r \not\sim \begin{pmatrix} 1 & 0 \\ \star & \chi^{k-1} \end{pmatrix}.$$

This follows from (3.1) much in the same way as (5.2) does, because $\bar{\rho}_{g,\mathfrak{q}} \sim 1 \oplus \chi^{k-1}$, since the trace of every reduction is $\equiv 1 + \chi^{k-1} \pmod{\mathfrak{q}}$ and the determinant is $\equiv \chi^{k-1}$.

Theorem 6.5. *The restriction $r|_{D_p}$ of r to the decomposition group D_p at p is semi-simple; in fact, the order of $r(D_p)$ is prime to p.*

Let F be the residue field of $K_{g,\mathfrak{q}}$, so that $r_p = r|_{D_p}$ is a degree-2 F-representation of $D_p = \mathrm{Gal}(\bar{\mathbf{Q}}_p|\mathbf{Q}_p)$. As for the D-module M (5.4), it suffices to show that the F-plane which carries the representation r_p is the direct sum of two (distinct) r_p-stable F-lines, to conclude that r_p is semisimple.

We have seen that there is an r_p-stable F-line on which D_p acts trivally; it acts via χ^{k-1} on the quotient (6.4). I am grateful to Christophe Breuil for providing the following proof that there is also an r_p-stable F-line on which D_p acts via χ^{k-1}, the action on the quotient being trivial.

The basic idea is that r_p is (a reduction of) a *crystalline* representation, by Faltings–Jordan [9] and, as the weight k is $< p - 1$, Fontaine–Laffaille [10] implies that the sequence $0 \to \chi^0 \to r_p \to \chi^{k-1} \to 0$ must split. We omit the details.

In other words, (2.3)(i)–(iv) hold for r. This completes our sketch of the proof of (2.3) using the level-1 weight-k cuspidal eigenform g instead of the level-p weight-2 cuspidal eigenform f of §4. As we have seen in §2, (2.3)⇒(2.1).

7. The denominators of Bernoulli numbers

Let's conclude with Witt's unpublished proof, as presented in [2], about the prime divisors of the denominators of B_k. It implies in particular that the numerator of B_{p-1} is never divisible by p.

Theorem 7.1 (von Staudt–Clausen, 1840). *Let $k > 0$ be an even integer, and let l run through the primes. Then the number*

$$W_k = B_k + \sum_{l-1|k} \frac{1}{l} \tag{1}$$

is always an integer. For example, $W_{12} = B_{12} + \frac{1}{2} + \frac{1}{3} + \frac{1}{5} + \frac{1}{7} + \frac{1}{13} = 1$.

The British analyst Hardy says in his *Twelve lectures* (p. 11) that this theorem was rediscovered by Ramanujan "at a time of his life when he had hardly formed any definite concept of proof".

Proof (Witt). The idea is to show that W_k is a p-adic integer for every prime p. More precisely, we show that $B_k + p^{-1}$ (resp. B_k) is a p-adic integer if $p-1|k$ (resp. if not).

For an integer $n > 0$, let $S_k(n) = 0^k + 1^k + 2^k + \cdots + (n-1)^k$. Comparing the coefficients on the two sides of

$$1 + e^T + e^{2T} + \cdots + e^{(n-1)T} = \frac{e^{nT} - 1}{T} \frac{T}{e^T - 1},$$

we get $S_k(n) = \sum_{m \in [0,k]} \binom{k}{m} \frac{B_m}{k+1-m} n^{k+1-m}$. To recover B_k from $S_k(n)$, it is tempting to take the limit $\lim_{n \to 0} S_k(n)/n$, which doesn't make sense in the archimedean world. If, however, we make n run through the powers p^s of a fixed prime p, then, p-adically, $p^s \to 0$ as $s \to +\infty$, and

$$\lim_{s \to +\infty} S_k(p^s)/p^s = B_k. \tag{2}$$

Let us compare $S_k(p^{s+1})/p^{s+1}$ with $S_k(p^s)/p^s$. Every $j \in [0, p^{s+1}[$ can be uniquely written as $j = up^s + v$, where $u \in [0, p[$ and $v \in [0, p^s[$. Now,

$$\begin{aligned} S_k(p^{s+1}) &= \sum_{j \in [0, p^{s+1}[} j^k = \sum_{u \in [0,p[} \sum_{v \in [0,p^s[} (up^s + v)^k \\ &\equiv p\left(\sum_v v^k\right) + kp^s\left(\sum_u u \sum_v v^{k-1}\right) \end{aligned}$$

modulo p^{2s}, by the binomial theorem. As $\sum_v v^k = S_k(p^s)$ and $2\sum_u u = p(p-1) \equiv 0 \pmod{p}$, we get

$$S_k(p^{s+1}) \equiv pS_k(p^s) \pmod{p^{s+1}},$$

where, for $p = 2$, the fact that k is even has been used. Dividing throughout by p^{s+1}, this can be expressed by saying that

$$\frac{S_k(p^{s+1})}{p^{s+1}} - \frac{S_k(p^s)}{p^s} \in \mathbf{Z}_{(p)}$$

is a p-adic integer, and therefore

$$\frac{S_k(p^r)}{p^r} - \frac{S_k(p^s)}{p^s} \in \mathbf{Z}_{(p)}$$

for any two integers $r > 0$, $s > 0$, since $\mathbf{Z}_{(p)}$ is a subring of $\mathbf{Q}$. Fixing $s = 1$ and letting $r \to +\infty$, we see that $B_k - S_k(p)/p \in \mathbf{Z}_{(p)}$, in view of (2). We need

Theorem 7.2. $S_k(p) = \sum_{j\in[1,p[} j^k \equiv \begin{cases} -1 \pmod{p} & \textit{if } p-1|k \\ 0 \pmod{p} & \textit{otherwise.} \end{cases}$

This is clear if $p - 1|k$. Suppose not, and let g be a generator of $(\mathbf{Z}/p\mathbf{Z})^\times$. We have $g^k - 1 \not\equiv 0$, whereas

$$(g^k - 1)\left(\sum_{j\in[1,p[} j^k\right) \equiv (g^k - 1)\left(\sum_{t\in[0,p-1[} g^{tk}\right) \equiv g^{(p-1)k} - 1 \equiv 0.$$

It follows that $B_k + p^{-1} \in \mathbf{Z}_{(p)}$ if $p - 1|k$ and $B_k \in \mathbf{Z}_{(p)}$ otherwise. In either case, the number W_k (1), which can be written as

$$W_k = \begin{cases} (B_k + p^{-1}) + \sum_{l\neq p} l^{-1} & \text{if } p-1|k \\ (B_k) + \sum_l l^{-1} & \text{otherwise,} \end{cases}$$

(where l runs through the primes for which $l - 1|k$) turns out to be a p-adic integer for every prime p. Hence $W_k \in \mathbf{Z}$, as claimed.

John Coates remarked at the workshop that the analogue of (7.1) for a totally real number field F (other than $\mathbf{Q}$) is an open problem; even a weak analogue would imply Leopoldt's conjecture for F.

References

[1] J. Bellaïche, À propos d'un lemme de Ribet, *Rend. Sem. Mat. Univ. Padova*, **109** (2003) 45–62.

[2] J. W. S. Cassels, Local fields, Cambridge University Press, Cambridge (1986), xiv+360 pp.

[3] J. W. S. Cassels and A. Fröhlich, Algebraic Number Theory, Academic Press, London (1967), xviii+366 pp.

[4] C. W. Curtis and I. Reiner, Representation theory of finite groups and associative algebras, Interscience, New York (1962), xiv+685 pp.

[5] P. Deligne, Formes modulaires et représentations l-adiques, *Sém. Bourbaki, Exp. 355, LNM*, **179**, Springer, Berlin (1969) 139–172.

[6] P. Deligne and M. Rapoport, Les schémas de modules de courbes elliptiques, Modular functions of one variable II, LNM 349, Springer, Berlin (1973) 143–316.

[7] P. Deligne and J.-P. Serre, Formes modulaires de poids 1, *Ann. sci. École norm. sup. (4)*, **7** (1975) 507–530.

[8] F. Diamond and J. Shurman, A first course in modular forms, Springer-Verlag, New York (2005), xvi+436 pp.

[9] G. Faltings and B. W. Jordan, Crystalline cohomology and GL(2, **Q**), *Israel J. Math.*, **90** (1995) no. 1–3, 1–66.

[10] J.-M. Fontaine and G. Laffaille, Construction de représentations p-adiques, *Ann. sci. École norm. sup. (4)*, **15** (1982) no. 4, 547–608.

[11] C. Khare, Notes on Ribet's converse to Herbrand, Cyclotomic fields, Bhaskaracharya Pratishthana, Poona (2000) 273–284.

[12] M. Raynaud, Schémas en groupes de type $(p, p, \dots, p)$, *Bull. Soc. Math. France*, **102** (1974) 241–280.

[13] K. A. Ribet, A modular construction of unramified p-extensions of $\mathbf{Q}(\mu_p)$, *Invent. Math.*, **34** (1976) no. 3, 151–162.

[14] K. A. Ribet, Bernoulli numbers and ideal classes, *Gaz. math.*, **118** (2008) 42–49.

[15] A. Saikia, Ribet's construction of a suitable cusp eigenform. This volume.

[16] J.-P. Serre, Une interprétation des congruences relatives à la fonction τ de Ramanujan. Sém. DPP, Fasc. 1, Exp. 14, 17 pp. Secrétariat mathématique, Paris (1969).

[17] J.-P. Serre, Abelian l-adic representations and elliptic curves, W. A. Benjamin, New York-Amsterdam (1968), xvi+177 pp.

[18] L. C. Washington, Introduction to cyclotomic fields Springer-Verlag, New York (1997), xiv+487 pp.

Guwahati Workshop on Iwasawa Theory of Totally Real Fields No. 12, 2010, pp. 83–95.

Fitting Ideals

F. A. E. Nuccio Mortarino Majno di Capriglio

Ruprecht-Karls-Universität Heidelberg, Im Neuenheimer Feld 288, 69120 Heidelberg, Deutschland
e-mail: nuccio@mathi.uni-heidelberg.de

1. General results on Fitting ideals

By "ring" we mean a commutative ring with 1. We fix such a ring and we call it R. If M is a finitely generated R-module and $\underline{x}$ is a set of generators of M (we will assume all such sets to be *ordered*), we indicate by $|\underline{x}|$ the number of elements it contains; an element $r = (r_1, \ldots, r_{|\underline{x}|}) \in R^{|\underline{x}|}$ is called a relation among the generators in $\underline{x}$ if

$$\sum_{i=1}^{|\underline{x}|} r_i x_i = 0 \in M.$$

Definition 1. *Let M be a finitely generated R-module and let $\underline{x}$ be a set of generators of M. A matrix X is said to be a matrix of relations for $\underline{x}$ if*

1. $X \in \mathcal{M}_{q\times|\underline{x}|}(R)$ *with* $q \geq |\underline{x}|$;
2. $X \cdot \underline{x} = \underline{0}$, i.e. *every row is a relation among the x_i's.*

We are now ready to define the main object of our study. If X is a $n \times m$-matrix (with $n \geq m$), we define its determinantal ideal $\mathrm{DetId}(X) \subseteq R$ to be the ideal of R generated by all $m \times m$-minors of X.

Definition 2. *Let M be a finitely generated R-module. We define the R-Fitting ideal of M to be*

$$\mathrm{Fitt}_R(M) = \sum_X \mathrm{DetId}(X)$$

where X runs through all matrices of relations for all sets of generators of M.

Remark. Observe that *a priori* the sum in the definition is extended to infinitely many ideals. Moreover, we want to stress that our convention that a matrix of relations has always more rows than columns (these being as many as the generators) is crucial for the definition of the determinantal ideal. Indeed, suppose that we allow every matrix whose rows are relations among a set of generators to be a *matrix of relations* and that we define its determinantal ideal to be the ideal generated by all minors of largest possible rank. Then if $\underline{x}$ is a set of generators, let $\underline{x}' = \{x_1, x_1, x_2, \ldots, x_{|\underline{x}|}\}$ be the old set of generators with one entry doubled – say the first. Then

$$(-1, 1, 0, \ldots, 0)$$

would be a matrix of relations, and $\mathrm{DetId}(X) = R$.

The definition of Fitting ideal given above is practically useless for any application: but it may be of some use in certain proofs. Thus, before passing to concrete examples, we want to show that in the definition it would have been enough to sum over all matrices of relations of *one fixed set of generators*. Indeed we have:

Proposition 3. *Let $\underline{x}$ be a fixed, finite set of generators of the finitely generated module M. Then*

$$\mathrm{Fitt}_R(M) = \sum_X \mathrm{DetId}(X)$$

where X runs through all matrices of relations for $\underline{x}$.

Proof. Only for the proof, for every set of generators $\underline{y}$ of M, set

$$A(\underline{y}) := \sum_Y \mathrm{DetId}(Y)$$

for Y running, as above, through all matrices of relations for $\underline{y}$. We need to show that for every set of generators $\underline{y}$, the equality $A(\underline{x}) = A(\underline{y})$ holds: indeed, by definition, $\mathrm{Fitt}_R(M) = \sum_{\underline{y}} A(\underline{y})$ and if they all coincide this sum reduces to $A(\underline{x})$.

Let then $\underline{y}$ be any set of generators and set $\underline{xy} := \{x_1, \ldots, x_{|\underline{x}|}, y_1, \ldots, y_{|\underline{y}|}\}$: clearly $|\underline{xy}| = |\underline{x}| + |\underline{y}|$. Since the x_i's generate M, we can find a relation

$$c_{k1}x_1 + c_{k2}x_2 + \cdots c_{k|\underline{x}|}x_{|\underline{x}|} + y_k = 0,$$

for suitable $c_{ki} \in R$, for every $1 \leq k \leq |\underline{y}|$. Let now $X = (a_{ij})$ be any $p \times |\underline{x}|$-matrix appearing in the sum $A(\underline{x})$ and define a new matrix, having $|\underline{x}| + |\underline{y}|$ columns,

$$\tilde{X} = \begin{pmatrix} c_{11} & \ldots & c_{1|\underline{x}|} & 1 & \ldots & 0 \\ \vdots & \ddots & \vdots & \vdots & \ddots & \vdots \\ c_{|\underline{y}|1} & \ldots & c_{|\underline{y}||\underline{x}|} & 0 & \ldots & 1 \\ a_{11} & \ldots & a_{1|\underline{x}|} & 0 & \ldots & 0 \\ \vdots & \ddots & \vdots & \vdots & \ddots & \vdots \\ a_{p1} & \ldots & a_{p|\underline{x}|} & 0 & \ldots & 0 \end{pmatrix}.$$

It is clearly a matrix of relations for $\underline{xy}$ (it has $p + |\underline{y}| \geq |\underline{x}| + |\underline{y}|$ rows) and every $|\underline{x}| \times |\underline{x}|$-minor of X also appears as $|\underline{xy}| \times |\underline{xy}|$-minor of $\tilde{X}$, showing the inclusion $A(\underline{x}) \subseteq A(\underline{xy})$.

Let now Y be a matrix of relations for $\underline{xy}$; define a new matrix

$$\tilde{Y} = \left(\begin{array}{cccc|c} c_{11} & c_{12} & \cdots & c_{1|\underline{x}|} & \\ \vdots & \vdots & \ddots & \vdots & \mathbb{I}_{|\underline{y}|} \\ c_{|\underline{y}|1} & c_{|\underline{y}|2} & \cdots & c_{|\underline{y}||\underline{x}|} & \\ \hline & & Y & & \end{array}\right)$$

where $\mathbb{I}_{|\underline{y}|}$ is the $|\underline{y}| \times |\underline{y}|$ identity matrix. Then $\mathrm{DetId}(\tilde{Y}) \supseteq \mathrm{DetId}(Y)$ (just observe that Y and $\tilde{Y}$ have the same number of columns and every $|\underline{xy}| \times |\underline{xy}|$-minor of Y is also a $|\underline{xy}| \times |\underline{xy}|$-minor of $\tilde{Y}$) and $\tilde{Y}$ is row equivalent to

$$\tilde{Y}' = \left(\begin{array}{cccc|c} c_{11} & c_{12} & \cdots & c_{1|\underline{x}|} & \\ \vdots & \vdots & \ddots & \vdots & \mathbb{I}_{|\underline{y}|} \\ c_{|\underline{y}|1} & c_{|\underline{y}|2} & \cdots & c_{|\underline{y}||\underline{x}|} & \\ \hline & & X & & 0 \end{array}\right)$$

(for this, use the identity matrix in the upper right corner of Y to make every element of the lower right corner trivial by means of elementary row operations); since the determinantal ideals of similar matrices clearly coincide, we find that $\mathrm{DetId}(Y) \subseteq \mathrm{DetId}(\tilde{Y}) = \mathrm{DetId}(\tilde{Y}') = \mathrm{DetId}(X)\mathrm{DetId}(\mathbb{I}_{|\underline{y}|}) = \mathrm{DetId}(X)$. Moreover, since the rows of $\tilde{Y}$ are relations among the elements of $\underline{xy}$, the same holds for the rows of $\tilde{Y}'$, so X is a matrix of relations for $\underline{x}$ (since Y was a matrix of relations, X has more rows than columns). This shows the inclusion $A(\underline{xy}) \subseteq A(\underline{x})$, and repeating the same argument with $\underline{y}$ at the place of $\underline{x}$ shows that $A(\underline{y}) = A(\underline{xy}) = A(\underline{x})$, as we needed. □

Remark. Observe that it is enough to restrict the computation to *square* $|\underline{x}| \times |\underline{x}|$-matrices. Indeed, substituting a matrix of relations X with all its $|\underline{x}| \times |\underline{x}|$-submatrices in the sum defining $\mathrm{DetId}(X)$ gives the same elements in the determinantal ideal.

Example. Let $M = R^a$ for some $a \geq 1$. A set of generators for M is $\underline{e}$, where e_i is the vector such that $(e_i)_j = \delta_{ij}$. For every relation $(r_1, \dots, r_a) \in R^a$,

$$\sum_{i=1}^{a} r_i e_i = 0 \Longrightarrow r_i = 0 \quad \forall\, i.$$

Therefore there are no non-trivial relations and, in particular, the only matrix of relations for $\underline{e}$ is the 0 matrix. Accordingly, $\mathrm{Fitt}_R(R^a) = 0$.

Example. Let p be a prime number, $R = \mathbb{Z}$ and let $M = \mathbb{Z}/p\mathbb{Z} \times \mathbb{Z}/p^2$. Then clearly $\mathrm{Ann}_{\mathbb{Z}}(M) = p^2\mathbb{Z}$. We want to compute $\mathrm{Fitt}_{\mathbb{Z}}(M)$. For this, let $\{(a, 0), (0, b)\}$ be a set of generators of M, where $p \nmid ab$. Then a matrix of relations for this set is of the form

$$X = \begin{pmatrix} \alpha_1 & \beta_1 \\ \alpha_2 & \beta_2 \\ \vdots & \vdots \\ \alpha_q & \beta_q \end{pmatrix}$$

with $v_p(\alpha_i) \geq 1$ and $v_p(\beta_i) \geq 2$, for some $q \geq 2$. Therefore, any 2×2-minor of X looks like

$$\det(X_{ij}) = \begin{vmatrix} \alpha_i & \beta_i \\ \alpha_j & \beta_j \end{vmatrix}$$

and $v_p(\det(X_{ij}) = v_p(\alpha_i\beta_j - \alpha_i\beta_i) \geq 3$, showing via Proposition 3 that $\mathrm{Fitt}_{\mathbb{Z}}(M) = p^3\mathbb{Z}$. If we allowed "matrices of relations" having only one row, (p, p^2) would have been one such a matrix, generating the ideal $p^2\mathbb{Z}$. In this sense, this invariant is "finer" than the usual annihilator.

Remark. In his book [Nor76] Northcott defines the i-th Fitting ideal for every natural number $i \geq 0$. What we have defined is then the 0-th Fitting ideal, and is enough for the purposes of Wiles' proof.

We now gather the main elementary results on Fitting ideals.

Lemma 4. *Let $I \subseteq R$ be an ideal. Then* $\mathrm{Fitt}_R(R/I) = I$.

Proof. A generator for R/I is 1 and the general matrix of relations for it is

$$\begin{pmatrix} \alpha_1 \\ \dots \\ \alpha_q \end{pmatrix}$$

such that $\alpha_i \cdot 1 = 0 \in R/I$, i.e. such that $\alpha_i \in I$ for all i's. □

For the rest of this section M will denote a finitely generated R-module.

Lemma 5. *Let M' be a quotient of M and let $\pi : M \twoheadrightarrow M'$ be a surjection: then* $\mathrm{Fitt}_R(M) \subseteq \mathrm{Fitt}_R(M')$.

Proof. Let $\underline{x}$ be a set of generators for M; then $\underline{\pi(x)} = \{\pi(x)_1, \dots, \pi(x)_{|\underline{x}|}\}$ is a set of generators for M'. If $\{r_1, \dots, r_{|\underline{x}|}\}$ is a relation for $\underline{x}$, it is also one for $\underline{\pi(x)}$. Therefore any matrix of relations for $\underline{x}$ is also a matrix of relations for $\underline{\pi(x)}$, showing the desired inclusion. □

Lemma 6. *If $M \cong M_1 \times M_2$, then* $\mathrm{Fitt}_R(M) = \mathrm{Fitt}_R(M_1) \cdot \mathrm{Fitt}_R(M_2)$.

Proof. On one side, any two matrices X, Y of relations for M_1 and M_2, respectively, give a matrix of relations

$$\begin{pmatrix} X & 0 \\ 0 & Y \end{pmatrix}$$

for $M_1 \times M_2$. On the other side, choosing as set of generators for $M_1 \times M_2$ one of the form $\underline{xy} = \{x_1, \dots, x_{|\underline{x}|}, y_1, \dots, y_{|\underline{y}|}\}$ where $\underline{x}$ is a set of generators for M_1 and $\underline{y}$ is one for M_2, we can form from any matrix of relations for $\overline{xy}$ two matrices of relations for $\underline{x}$ and $\underline{y}$. These matrices verify clearly $\mathrm{DetId}(X)\mathrm{DetId}(Y) = \mathrm{DetId}(XY)$, and our statement follows. □

Combining Lemma 6 and Lemma 4 we get

Corollary 7. *If* $M = R/\mathfrak{a}_1 \times \cdots \times R/\mathfrak{a}_n$ *for ideals* $\mathfrak{a}_i \subseteq R$, *then* $\mathrm{Fitt}_R(M) = \mathfrak{a}_1 \cdots \mathfrak{a}_n$. □

Lemma 8. *If M can be generated by n elements, then*

$$\mathrm{Ann}_R(M)^n \subseteq \mathrm{Fitt}_R(M) \subseteq \mathrm{Ann}_R(M).$$

Proof. First of all, let $\underline{x}$ be a set of generators with $|\underline{x}| = n$. Then, if $c_i \in \mathrm{Ann}_R(M)$ for $1 \leq i \leq n$, the matrix

$$C = \begin{pmatrix} c_1 & \dots & 0 \\ \vdots & \ddots & \vdots \\ 0 & \dots & c_n \end{pmatrix}$$

is a matrix of relations for $\underline{x}$ and $c_1 \cdots c_n = \det(C) \in \mathrm{Fitt}_R(M)$, so $\mathrm{Ann}_R(M)^n \subseteq \mathrm{Fitt}_R(M)$.

On the other hand, let X be a square $n \times n$-matrix of relations for $\underline{x}$: let X° be its *adjugate* matrix[1]. Then multiplying the equation

$$X \cdot \underline{x} = \underline{0}$$

by X° shows that $\det(X) \in \mathrm{Ann}_R(M)$. Since, as we observed, $\mathrm{Fitt}_R(M)$ is generated by these determinants, we find $\mathrm{Fitt}_R(M) \subseteq \mathrm{Ann}_R(M)$. □

Remark. This Lemma gives another proof that $\mathrm{Fitt}_R(R/I) = I$, since in that case $n = 1$ and $\mathrm{Ann}_R(M) = I$.

During the school, Otmar Venjakob observed that since M admits n generators, there is a surjection

$$(R/\mathrm{Ann}_R(M))^n \twoheadrightarrow M$$

and applying Lemma 5 gives the first inclusion.

Recall that a R-module M is said to be faithful if $\mathrm{Ann}_R(M) = 0$. Thus the above Lemma gives

Corollary 9. *If M is a faithful R-module, then* $\mathrm{Fitt}_R(M) = 0$. □

Lemma 10. *If* $I \subseteq R$ *is any ideal of R, then*

$$\mathrm{Fitt}_{R/I}(M/IM) = \mathrm{Fitt}_R(M) + I \subseteq R/I$$

Proof. It is clear that if $\underline{x}$ is a set of generators for M, than its image under the canonical projection is a set of generators for M/I, and that any relation for the first set is also a relation for the second, giving $\mathrm{Fitt}_R(M) + I \subseteq \mathrm{Fitt}_{R/I}(M/IM)$.

[1] The adjugate matrix X° of a square $n \times n$-matrix X with coefficients in a ring R is a matrix such that $XX^\circ = \det(X)\mathbb{I}_n$. It may be defined as the matrix whose (i, j)-entry is the (j, i)-cofactor of X.

If now $[X] = ([\xi_{ij}])$ is a matrix of relations for the set $[\underline{x}]$ of generators of M/IM (we denote by $[\cdot]$ the class $\cdot + I$), we can find elements $\alpha_{ij} \in I$ such that for all $1 \leq i \leq |\underline{x}|$ we have

$$\sum_{j=1}^{|\underline{x}|} \xi_{ij} x_j = \sum_{j=1}^{|\underline{x}|} \alpha_{ij} x_j$$

because every element of IM can be written as a finite sum of the generators of M with coefficients in I. The matrix $\tilde{X}$ whose i-th row is

$$(\xi_{i1} - \alpha_{i1}, \ldots, \xi_{i|\underline{x}|} - \alpha_{i|\underline{x}|})$$

is a matrix of relations for M and $X \equiv \tilde{X} \pmod{I}$. Therefore $\operatorname{DetId}([X]) = \operatorname{DetId}([\tilde{X}]) = \operatorname{DetId}(\tilde{X}) + I \subseteq \operatorname{Fitt}_R(M) + I$. This gives the other inclusion. □

Lemma 11. *If* $0 \longrightarrow M_1 \longrightarrow M_2 \longrightarrow M_3 \longrightarrow 0$ *is an exact sequence of finitely generated R-modules, then*

$$\operatorname{Fitt}_R(M_1) \cdot \operatorname{Fitt}_R(M_3) \subseteq \operatorname{Fitt}_R(M_2).$$

Proof. Without loss of generality, we may and will assume that $M_1 \subseteq M_2$ and that the first map above is the natural inclusion. Let now $[\underline{x}]$ be a set of generators for M_3 and let $\underline{x}$ be a lifting of it to M_2. Let $\underline{y}$ be a set of generators for $M_1 \subseteq M_2$. Then $\underline{xy}$ is a set of generators for M_2.

Let $X \in \mathcal{M}_{p\times|\underline{x}|}(R)$, $Y \in \mathcal{M}_{q\times|\underline{y}|}(R)$ be matrices of relations for $[\underline{x}]$ and $\underline{y}$, respectively, and let $C = (-c_{ij}) \in \mathcal{M}_{p\times|\underline{y}|}(R)$, where

$$X \cdot \underline{x} = \begin{pmatrix} \sum_{i=1}^{|\underline{y}|} c_{1i} y_i \\ \sum_{i=1}^{|\underline{y}|} c_{2i} y_i \\ \vdots \\ \sum_{i=1}^{|\underline{y}|} c_{pi} y_i \end{pmatrix}.$$

Then the matrix

$$Z = \begin{pmatrix} X & C \\ 0 & Y \end{pmatrix}$$

is a $(p+q) \times |\underline{xy}|$-matrix of relations for M_2 and $\operatorname{DetId}(X) \cdot \operatorname{DetId}(Y) \subseteq \operatorname{DetId}(Z)$ since every minor of Z is of the form

$$\begin{vmatrix} M_X & M_C \\ 0 & M_Y \end{vmatrix} = \begin{vmatrix} M_X & 0 \\ 0 & M_Y \end{vmatrix}$$

where M_X is a submatrix of X and M_Y is a submatrix of Y. □

2. Applications to Wiles' proof

In this section we focus on the applications of the theory of Fitting ideals developed up to now to Wiles' proof. In a first part we will start by assuming that not only M is finitely generated, but that is in fact finitely presented; and we eventually specialize to the case $R = \mathcal{O}[[T]]$ where $\mathcal{O}$ is the ring of integers of a p-adic field.

2.1 *Finite presentations*

First of all, recall that an R-module M is said to be *finitely presented* if there is an exact sequence of R-modules

$$R^a \xrightarrow{h} R^b \xrightarrow{\varphi} M \to 0$$

for suitable $a, b \in \mathbb{N}$. Clearly, every finitely presented module is finitely generated, since it can be generated by $\varphi(e_i)$ for $1 \le i \le b$. If R is noetherian, the converse is also true: indeed, if M is finitely generated, it is a quotient of a finitely generated free module and there is an exact sequence

$$0 \to S \xrightarrow{\psi} R^b \xrightarrow{\varphi} M \to 0$$

for some S and for some $b \in \mathbb{N}$. Using that R is noetherian, S should also be finitely generated, say a quotient of R^a by a module T: we find

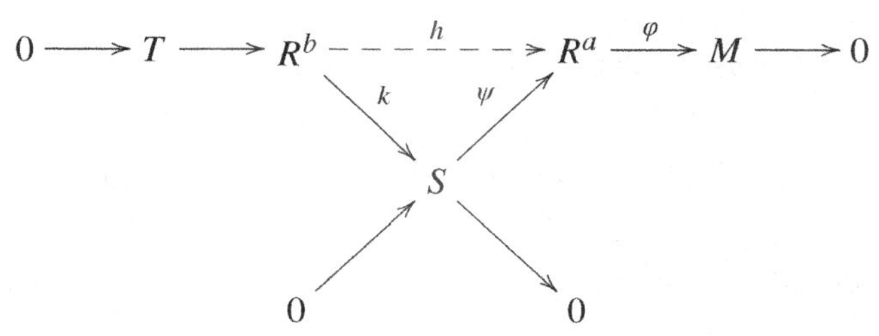

where $h := \psi \circ k$. We can therefore extract from the above diagram the sequence

$$R^a \xrightarrow{h} R^b \xrightarrow{\varphi} M \to 0$$

that is exact since $\mathrm{Ker}(\varphi) = \mathrm{Im}(\psi) = \mathrm{Im}(h)$.

Now we present some results that are true for general finitely presented R-modules. Thus, assuming noetherianity reduces this hypothesis to M being finitely generated.

From now on we fix an R-module M and we assume it is finitely presented.

Proposition 12. *Let* $R^a \xrightarrow{h} R^b \xrightarrow{\varphi} M \to 0$ *be a finite presentation of* M *and let* $H \in \mathcal{M}_{b\times a}(R)$ *be a matrix attached to* h. *If* $a \ge b$, *then* $\mathrm{Fitt}_R(M) = \mathrm{DetId}(H^{tr})$ *and if* $a < b$ *then* $\mathrm{Fitt}_R(M) = 0$.

Proof. The finite presentation

$$R^a \xrightarrow{h} R^b \xrightarrow{\varphi} M \to 0$$

shows that a set of generators of M is $\{\varphi(e_1), \dots, \varphi(e_b)\}$ and that the relations among these elements are precisely the kernel of φ, namely $\operatorname{Im}(h)$, that is spanned by the columns of H. By Proposition 3, to compute $\operatorname{Fitt}_R(M)$ it suffices to consider matrices whose rows are relations among these generators.

If $a \geq b$, then H^{tr} is a matrix of relations; moreover, if X is any other matrix of relations, then all its sumbatrices are linear combinations of rows of H^{tr} and their determinant coincide with the determinant of those rows. Therefore $\operatorname{DetId}(X) \subseteq \operatorname{DetId}(H^{\mathrm{tr}})$, showing the desired equality.

If $a < b$, then every matrix of relations $X \in \mathcal{M}_{b\times b}(R)$ has at least two linearly dependent rows, since they belong to $\operatorname{Im}(h)$ and this module has R-rank $a < b$. Therefore $\operatorname{DetId}(X) = 0$ for all such matrices, showing $\operatorname{Fitt}_R(M) = 0$. □

Lemma 13. *If I is a finitely generated ideal, then*

$$\operatorname{Fitt}_R(M) \subseteq \operatorname{Fitt}_R(M/IM) \subseteq (\operatorname{Fitt}_R(M), I) \subseteq R.$$

Proof. The first inclusion is Lemma 5. For the second, if

$$R^a \xrightarrow{h} R^b \to M \to 0$$

is a presentation of M and if $I = \langle \alpha_1, \dots, \alpha_n \rangle$, then a presentation of M/IM is

$$R^{a+nb} \xrightarrow{h^*} R^b \to M/IM \to 0$$

where

$$h^*(w_1, \dots, w_a, v_{1,1}, \dots, v_{1,b}, \dots, v_{n,1}, \dots, v_{n,b})$$
$$= h(w_1, \dots, w_a) + \sum_{j=1}^{n} (\alpha_j v_{j,1}, \dots, \alpha_j v_{j,b}).$$

A matrix associated to h^* is

$$H^* = \begin{pmatrix} H & 0 & \dots & 0 \\ 0 & A & \dots & 0 \\ \vdots & \vdots & \ddots & \vdots \\ 0 & 0 & \dots & A \end{pmatrix}$$

where H is a matrix associated to h and A is the diagonal matrix

$$A = \begin{pmatrix} \alpha_1 & 0 & \dots & 0 \\ 0 & \alpha_2 & \dots & 0 \\ \vdots & \vdots & \ddots & \vdots \\ 0 & 0 & \dots & \alpha_n \end{pmatrix}.$$

Any non-trivial $b \times b$ minor of H^* is either a $b \times b$ minor of H or a linear combination of suitable multiples of some α_j's, and hence we get the second inclusion. Observe that the above proof works both if $a \geq b$ and if $a < b$. □

Combining this Lemma with Corollary 9 we get

Corollary 14. *If $I \subseteq R$ is a finitely generated ideal and M is a faithful R-module, then* $\mathrm{Fitt}_R(M/IM) \subseteq I$. □

2.2 *Iwasawa algebras*

Next we focus on applications of the theory of Fitting ideal in Wiles' proof of the Iwasawa Conjecture. We assume for this that $\mathcal{O}$ is the ring of integers of a p-adic field and we let Λ denote the formal power series ring $\Lambda = \mathcal{O}[[T]]$: we refer the reader to chapters 7, 13 and 15 of [Was97] for basic facts about Λ-modules. This is a complete, local, noetherian ring: let $\mathfrak{m}$ denote its maximal ideal, that is generated (for instance) by π and T where π is a uniformizer of $\mathcal{O}$. Since it is noetherian, we can combine general results of Section 1 with those of §2.1.

If $\mathfrak{p}_1, \dots, \mathfrak{p}_k$ are prime ideals of height 1 in Λ and $r, e_1, \dots, e_k$ are non-negative integers, define the finitely generated Λ-module $E(r, \mathfrak{p}_1^{e_1}, \dots, \mathfrak{p}_k^{e_k})$ to be

$$E(r, \mathfrak{p}_1^{e_1}, \dots, \mathfrak{p}_k^{e_k}) = \Lambda^r \oplus \bigoplus_{i=1}^{k} \Lambda/\mathfrak{p}_i^{e_i} :$$

These are called *elementary* Λ-modules: they are classified (up to isomorphism) by the above set of prime ideals and integers and every finitely generated Λ-module is pseudo-isomorphic to exactly one of these modules – a pseudo-isomorphism being a Λ-homomorphism with finite kernel and cokernel.

For a finitely generated *torsion* module M, $r = 0$ and a coarser (but often important) invariant than the entire set above is the so-called characteristic polynomial, that is the unique distinguished[2] polynomial $f_M(T)$ such that

$$f_M(T) \cdot \Lambda = \prod_{i=1}^{k} \mathfrak{p}_i^{e_i};$$

since this definition is easily seen to be multiplicative in exact sequences and since finite modules are pseudo-isomorphic to the trivial module, this is an invariant under pseudo-isomorphism. Observe that prime ideals of height 1 are all principal, generated either by a uniformizer of $\mathcal{O}$ or by irreducible distinguished polynomial. We start with the following

Lemma 15. *Let $N = E(0, \mathfrak{p}_1^{e_1}, \dots, \mathfrak{p}_k^{e_k})$ be an elementary torsion Λ-module. Then*

(1) $f_N(T) \cdot \Lambda = \mathrm{Fitt}_\Lambda(N)$.
(2) *If none of the $\mathfrak{p}_i$'s is $\pi \cdot \Lambda$, then*

$$\mathrm{Fitt}_\Lambda(N) = \mathrm{char}_\Lambda(\cdot T : N \to N) \cdot \Lambda.$$

where $\mathrm{char}_\Lambda(\cdot T : N \to N)$ *is the characteristic polynomial (say, in the variable X) of the multiplication by T seen as $\mathcal{O}$-linear endomorphisms of the finite free $\mathcal{O}$-module N,* evaluated at $X = T$.

[2] A power series is called a distinguished polynomial if it is either a constant in $\pi\mathcal{O}$ or it is a monic polynomial whose coefficients are all divisible by π but the leading one.

Remark. Given a non-constant distinguished polynomial $g(T)$, the module $\Lambda/g(T)\Lambda$ is $\mathcal{O}$-torsion free; thus, if $\mathfrak{p}_i \neq \pi\Lambda$ for all i's, then N is finitely generated and torsion-free over the principal ideal domain $\mathcal{O}$, hence free of finite rank. Moreover, despite its name, we will always denote the characteristic polynomial of a Λ-module M by f_M, leaving the notation $\mathrm{char}_\Lambda(\cdot)$ for the usual meaning of the characteristic polynomial of a linear map (see Corollary 16 below for instances in which these two notions coincide).

Proof. The equality in (1) is clear after Corollary 7 and the definition of elementary module.

For (2) without loss of generality we can assume that $N = \Lambda/\mathfrak{p}^e$ with $\mathfrak{p} = g(T)\Lambda$ where $g(T)$ is a non-constant distinguished polynomial. Then $\mathfrak{p}^e = g(T)^e\Lambda$ and $g(T)^e$ is still distinguished, say of degree λ: let g_i be the coefficient of T^i in $g(T)^e$ for $0 \leq i \leq \lambda - 1$. Let $\{1, T, \ldots, T^{\lambda-1}\}$ be a basis of N: the matrix attached to the multiplication by T in this basis is

$$\begin{pmatrix} 0 & 0 & 0 & 0 & -g_0 \\ 1 & 0 & 0 & 0 & -g_1 \\ \vdots & 1 & 0 & 0 & -g_2 \\ \vdots & \vdots & \ddots & \vdots & \vdots \\ 0 & 0 & 0 & 1 & -g_{\lambda-1} \end{pmatrix}$$

and the characteristic polynomial is then

$$\begin{vmatrix} -X & 0 & 0 & 0 & -g_0 \\ 1 & -X & 0 & 0 & -g_1 \\ \vdots & 1 & -X & 0 & -g_2 \\ \vdots & \vdots & \ddots & \ddots & \vdots \\ 0 & 0 & 0 & 1 & -g_{\lambda-1} - X \end{vmatrix}$$

Developing the determinant along the first line shows, by induction on λ, that this characteristic polynomial is precisely

$$\sum_{i=0}^{\lambda-1} g_i X^i + X^\lambda = g(X)^e.$$

Using (1) and evaluating at $X = T$ yields (2), since $f_{\Lambda/g(T)^e\Lambda} = g(T)^e$. □

Corollary 16. *Let N be a finitely generated torsion Λ-module having no $\mathcal{O}$-torsion. Then N is a finite free $\mathcal{O}$-module and*

$$f_N(T) \cdot \Lambda = \mathrm{char}_\Lambda(\cdot T : N \to N) = \mathrm{Fitt}_\Lambda(E_N)$$

where E_N is the elementary module pseudo-isomorphic to N.

Proof. The freeness of N follows, as before, from $\mathcal{O}$ being a principal ideal domain; observe also that the elementary module attached to a module without $\mathcal{O}$-torsion does not have $\pi \cdot \Lambda$ in its set of invariants. Thanks to the above Lemma, the equality follows from $f_N(T) = f_{E_N}(T)$, which is clear since, by definition, the characteristic polynomial of a Λ-module depends only on its pseudo-isomorphism class; and from $\mathrm{char}_\Lambda(\cdot T : N \to N) = \mathrm{char}_\Lambda(\cdot T : E_N \to E_N)$. This last relation may be proven, for instance, by observing that both are monic polynomials with coefficients in $\mathcal{O}$ of the same degree that have the same image in $\Lambda \otimes \overline{\mathbb{Q}_p}$. □

Proposition 17. *Let N be a finitely generated torsion Λ-module having no $\mathcal{O}$-torsion. Then*

$$f_N(T) \cdot \Lambda = \mathrm{Fitt}_\Lambda(N).$$

We start with the following Lemma:

Lemma 18 (Auslander-Buchsbaum resolution). *Let N be finitely generated torsion Λ-module that is free of finite rank as $\mathcal{O}$-module. Then*

$$\Lambda \otimes_{\mathcal{O}} N \xrightarrow{\gamma} \Lambda \otimes_{\mathcal{O}} N \xrightarrow{\varepsilon} N \to 0$$

is a finite presentation of N as Λ-module, where

$$\gamma(\lambda \otimes n) = T\lambda \otimes n - \lambda \otimes Tn$$

and

$$\varepsilon(\lambda \otimes n) = \lambda n.$$

Proof. The fact that ε is surjective is clear, as well as the fact that both γ and ε are Λ-homomorphisms and that $\varepsilon \circ \gamma = 0$. Moreover, if we prove that the sequence is exact, it is a presentation of N since the freeness of N implies that of $\Lambda \otimes N$.

Thus we need only to show that $\mathrm{Im}(\gamma) \supseteq \mathrm{Ker}(\varepsilon)$ and this would follow from

$$x = \gamma(\xi) + 1 \otimes \varepsilon(x) \qquad \forall\, x \in \Lambda \otimes_{\mathcal{O}} N \tag{1}$$

for some ξ depending on x. Since (1) is $\mathcal{O}$-linear, we can check it only on elements of the form $x = T^k \otimes n$ (use also that $\mathcal{O}[T]$ is dense in Λ and both γ, ε are continuous) for suitable $k \geq 0$ and $n \in N$: this we do by induction on k. If $k = 1$, $x = T \otimes n = T \otimes n - 1 \otimes Tn + 1 \otimes Tn = \gamma(1 \otimes n) + 1 \otimes \varepsilon(T \otimes n)$, as we wanted. Asssuming (1) up to $k - 1$, suppose $x = T^k \otimes n$: then

$$\begin{aligned} x &= TT^{k-1} \otimes n - T^{k-1} \otimes Tn + T^{k-1} \otimes Tn \overset{\text{induction}}{=} \\ &= T^{k-1}(T \otimes n - 1 \otimes Tn) + \gamma(\xi') + 1 \otimes \varepsilon(T^{k-1} \otimes Tn) \\ &= T^{k-1}\gamma(1 \otimes n) + \gamma(\xi') + 1 \otimes T^k n \\ &= \gamma(T^{k-1} \otimes n + \xi') + 1 \otimes \varepsilon(x). \end{aligned}$$

Setting $\xi = T^{k-1} \otimes n + \xi'$ we get (1). □

Remark. It is indeed easy to prove that γ is injective.

We can now prove Proposition 17:

Proof. If $\underline{x}$ is a $\mathcal{O}$-basis of N, the set $\{1 \otimes x_1, \dots, 1 \otimes x_{|\underline{x}|}\}$ is a Λ-basis of $\Lambda \otimes N$: moreover, $\gamma = \vartheta - \tau$ where

$$\vartheta : \lambda \otimes n \mapsto T\lambda \otimes n \qquad \tau : \lambda \otimes n \mapsto \lambda \otimes Tn$$

and therefore $\det \mathrm{Mat}(\gamma) = \det(\mathrm{Mat}(\vartheta) - \mathrm{Mat}(\tau))$, where we introduce – just along the proof – the notation $\mathrm{Mat}(\cdot)$ to denote the matrix attached to a linear map in the bases chosen above. Since ϑ maps every element of the Λ-basis $1 \otimes x_i$ to $T \otimes x_i$, and by definition of τ, we find

$$\mathrm{Mat}(\vartheta) = T \cdot \mathbb{I}_{|\underline{x}|} \qquad \tau = \cdot T : N \to N.$$

This shows that the characteristic polynomial of multiplication by T (evaluated at $X = T$) is the determinant of $\mathrm{Mat}(\gamma)$ and, by Proposition 12 applied to the Auslander-Buschsbaum resolution, we see that this determinant generates the Λ-Fitting ideal of N. Corollary 16 then gives the stated equality. □

Now that we have collected these results, let M be a finitely generated torsion Λ-module and assume that "$\mu = 0$", meaning[3] that π does not divide the characteristic polynomial $f_M(T)$ of M. Let $tor_{\mathcal{O}}(M) \subseteq M$ be the maximal $\mathcal{O}$-torsion submodule of M: it is a Λ-submodule of M and there is an exact sequence

$$0 \longrightarrow tor_{\mathcal{O}}(M) \longrightarrow M \longrightarrow N \longrightarrow 0 \tag{2}$$

where N satisfies the hypothesis of Corollary 16. Observe now that as an application of Lemma 11 we find

Corollary 19. *Suppose R is a local ring and $\mathfrak{m}$ is its maximal ideal. If M is an R-module of finite length, then*

$$\mathfrak{m}^{\mathrm{length}_R(M)} \subseteq \mathrm{Fitt}_R(M).$$

Proof. We claim that over a local ring every module of length 1 is isomorphic to $k := R/\mathfrak{m}$. Granting the claim, the Corollary follows from Lemma 4 and from Lemma 11, by induction on $\mathrm{length}_R(M)$.

We now prove the claim: observe that if M is a k-vector space endowed with the natural R-action induced by $R \twoheadrightarrow k$, then the notion of k-vector subspace and of R-submodule of M coincide. Therefore $\mathrm{length}_R(M) = \mathrm{length}_k(M) = \dim_k(M)$. The module M has length 1, and therefore in the chain of inclusions

$$M \supseteq \mathfrak{m}M \supseteq 0$$

one is an equality. By Nakayama Lemma, $M = \mathfrak{m}M \Rightarrow M = 0$, and this is absurd since $\mathrm{length}_R(0) = 0$: therefore $\mathfrak{m}M = 0$, M is a k-vector space and our previous discussion shows that it has dimension 1 over k. □

[3] Traditionally, μ is the exponent of $\pi \cdot \Lambda$ in the set of invariants of an elementary module.

Applying this to our situation, we get:

Proposition 20. *Let M be a finitely generated torsion Λ-module such that $\mu = 0$. Then*

$$f_M(T) \cdot (\pi, T)^{\mathrm{length}_\Lambda(tor_{\mathcal{O}}(M))} \subseteq \mathrm{Fitt}_\Lambda(M) \subseteq f_M(T) \cdot \Lambda.$$

Proof. Observe, first of all, that since $tor_{\mathcal{O}}(M)$ is finite, it has trivial characteristic polynomial and $f_M(T) = f_N(T)$. Now apply Lemma 11 to the above sequence (2). We find that

$$\mathrm{Fitt}_\Lambda(tor_{\mathcal{O}}(M)) \cdot \mathrm{Fitt}_\Lambda(N) \subseteq \mathrm{Fitt}_\Lambda(M) :$$

since $tor_{\mathcal{O}}(M)$ is of finite length and N has no $\mathcal{O}$-torsion, we can apply Corollary 19 and Corollary 16 (together with our first remark) to translate the above inclusion as $(\pi, T)^{\mathrm{lentgh}_\Lambda(tor_{\mathcal{O}}(M))} \cdot f_M(T) \subseteq \mathrm{Fitt}_\Lambda(M)$. The second inclusion just follows from Lemma 5 together with Corollary 16. □

References

[MW84] B. Mazur and A. Wiles, Class fields of abelian extensions of **Q**, *Invent. Math.*, **76** (1984) no. 2, 179–330.

[Nor76] D. G. Northcott, Finite free resolutions, Cambridge University Press, Cambridge (1976), Cambridge Tracts in Mathematics, No. 71.

[Was97] Lawrence C. Washington, Introduction to Cyclotomic Fields, second ed., Graduate Texts in Mathematics, vol. 83, Springer-Verlag, Berlin (1997).

[Wil90] A. Wiles, The Iwasawa conjecture for totally real fields, *Ann. of Math. (2)*, **131** (1990) no. 3, 493–540.

Guwahati Workshop on Iwasawa Theory of Totally Real Fields No. 12, 2010, pp. 97–112.

Deligne-Ribet's work on L-values

O. Venjakob

Mathematisches Institut, Universität Heidelberg, INF 288, D-69120 Heidelberg, Germany
e-mail: venjakob@mathi.uni-heidelberg.de

The aim of this series of 3 talks is to report on

I. Siegel's result on the *rationality* of the values $L(1-k,\varepsilon), k \geq 1$, of certain L-functions over totally real number fields to be defined later and
II. the *congruences* among those values which have been shown by Deligne and Ribet and which amount to the existence of *p-adic L-functions*.

1. Rationality

In the first lecture we concentrate on topic I. The main idea of Siegel which actually goes back already to Klingen consists of considering certain Eisenstein series the constant term of which are exactly the L-values above. Then he applies the following principle:

Let $f(z)$ be an $SL_2(\mathbb{Z})$-Eisenstein series, $z \in \mathbb{H}$, of weight k with Fourier series

$$f(z) = \sum_{n=0}^{\infty} a_n\, q^n, \qquad q = e^{2\pi i z}$$

and assume that $a_1, \ldots, a_k \in \mathbb{Q}$. Then it follows from the following result that also a_0 is rational.

Proposition 1.1 (Siegel). *There exist integers $c_{k,0}, \ldots, c_{k,r}$ with*

$$r := r(k) := \dim_{\mathbb{C}} M_k(SL_2(\mathbb{Z})) = \begin{cases} \left[\frac{k}{12}\right] + 1 & \text{if } k \not\equiv 2 \bmod 12, \\ \left[\frac{k}{12}\right] & \text{otherwise,} \end{cases}$$

such that

$$c_{k,0}\, a_0 + c_{k,1}\, a_1 + \cdots + c_{k,r}\, a_r = 0$$

and

$$c_{k,0} \neq 0.$$

Proof ([3, §5.2 Cor. 1], [12]). Let X be the compact Riemann surface

$$X = SL_2(\mathbb{Z})\backslash\mathbb{H} \cup \{\infty\} \underset{j}{\cong} \mathbb{P}^1$$

with meromorphic differentials $\Omega^1(X)$ and consider the $\mathbb{C}$-linear map

$$M_k(SL_2(\mathbb{Z})) \overset{\phi}{\hookrightarrow} \Omega^1(X)$$
$$f \mapsto w(f) := f\,T_k(z)dz = \frac{1}{2\pi i} f\,T_k(q)\frac{dq}{q}$$

where

$$T_k(z) := G_{14-k+12(r(k)-1)}\,\Delta^{-r} = c_{k,r}q^{-r} + \cdots + c_{k,0} + \cdots$$

is a weight $2-k$ modular function, holomorphic on $\mathbb{H}$, with integral Fourier coefficients $c_{k,r} \in \mathbb{Z}$. Here Δ is Ramanujan's Δ-function, a cusp form of weight 12, and G_k is the weight k-Eisenstein series with constant Fourier coefficient 1. Note that for the weights actually showing up in the T_k's, the q-expansions of the G_i are integral. It is easy to show that the image of ϕ has the basis $\omega_m := j^m dj$, $m = 0, \ldots, r-1$, where $j = \frac{G_4^3}{\Delta} = q^{-1} + \cdots$ is the modular j-function. The constant term of $T_k f$ is obviously

$$c_{k,0}\,a_0 + \cdots + c_{k,r}\,a_r$$

which equals the coefficient of q^{-1} in $\omega(f) = \frac{1}{2\pi i} T_k f \frac{dq}{q}$. But each $\omega_m = j^m dj = \frac{1}{m+1}\frac{dj^{m+1}}{dq}dq$ has trivial coefficient in q^{-1}, thus the first claim follows. The fact that $c_{k,0} \neq 0$ is an explicit calculation concerning G_i and Δ, see [3]. □

Remark 1. Siegel's argument in [13] is slightly different: he shows that for the linearly independent modular forms $1, \gamma_1, \ldots, \gamma_{r(k)}$ of weight 0 and k the vectors $1_t, (\gamma_1)_t, \ldots, (\gamma_{r(k)})_t$ of their first $t+1$ Fourier coefficients are still linearly independent (for $t = r(k)+1$). From this fact he is able to express a_0 as a linear combination of $a_1, \ldots, a_{r(k)}$. If $f(z)$ is a weight k Eisenstein series of $\Gamma(N)$ he uses the fact that $f(z)$ satisfies some algebraic relation with coefficients in $SL_2(\mathbb{Z})$-Eisenstein series up to a certain bounded weight ℓ. For the linearly independent modular forms $1, \gamma_1^{k_1}, \ldots, \gamma_{r(k_1)}^{k_1}, \ldots, \gamma_1^{k_d}, \ldots, \gamma_{r(k_d)}^{k_d}$ (γ_i^j of weight j and $k_d = \ell$) he determines an upper bound for t, the minimal integer such that the vectors $(1)_t, \ldots, (\gamma_{r(k_d)}^{k_d})_t$ stay linearly independent. Thereby he again manages to express a_0 as a rational linear combination of the $a_1, \ldots, a_t$ (with bounded denominators).

Since the coefficients $c_{k,j}$ are computable, we obtain explicit numerical formulae for the first Fourier coefficients.

Example 1. The classical Eisenstein series

$$\mathbb{E}_k(z) = \frac{(k-1)!}{2(2\pi i)^k} \sum_{(m,n)\in\mathbb{Z}^2\setminus\{0\}} (mz+n)^{-k},$$
$$= \frac{\zeta(1-k)}{2} + \sum_{n=1}^{\infty} \sigma_{k-1}(n)q^n$$

with $\sigma_k(n) := \sum_{0<d|n} d^k$ gives

$$\zeta(1-k) = -\frac{2}{c_{k,0}} \sum_{j=1}^{r(k)} \sigma_{k-1}(n) c_{k,j} \in \mathbb{Q}$$

for even integers $k > 2$. Here ζ denotes the Riemann zeta-function.

We now fix some notation for the rest of the talks:

K	totally real number field,
$r = [K : Q]$	
$\mathcal{O} = \mathcal{O}_K$	its ring of integers
$\mathfrak{D} = (\mathcal{O}^*)^{-1}$	different
$d_K = \mathcal{N}(\mathfrak{D})$	discriminant
$\mathcal{N} : K \to \mathbb{Q}$	(or $K \otimes_Q R \to R$ for any $\mathbb{Q}$-algebra R) the norm map.
$\mathfrak{f} \subset \mathcal{O}$	conductor
$\hat{K}$	the ring of finite adeles of K

$\alpha \in K$ is called *totally positive*, $\alpha \gg 0$, if $\sigma(\alpha) > 0$ for all embeddings $\sigma : K \hookrightarrow \mathbb{R}$.
$\mathbb{H}_K = \{\tau \in K \otimes \mathbb{C} | \operatorname{Im}(\tau) \gg \mathcal{O}\} \cong \mathbb{H}_{\mathbb{Q}}^r$

$$\Gamma_{00}(\mathfrak{f}) = \left\{ \begin{pmatrix} a & b \\ c & d \end{pmatrix} \in SL_2(K) | \begin{matrix} a, d \in 1 + \mathfrak{f} \\ b \in \mathfrak{D}^{-1},\ c \in \mathfrak{f}\mathfrak{D} \end{matrix} \right\}.$$

Let $\mathbb{E}_k(\tau)$, $\tau \in \mathbb{H}_K$, be an $SL_2(\mathcal{O})$-Eisenstein series of weight k. Then its pullback $f = (\Delta^* \mathbb{E}_k)(z)$ along the diagonal

$$\mathbb{H}_{\mathbb{Q}} \overset{\Delta}{\hookrightarrow} \mathbb{H}_K = \mathbb{H}_{\mathbb{Q}}^r$$
$$z \mapsto (z, \dots, z)$$

is an $SL_2(\mathbb{Z})$-Eisenstein series of weight kr with the same constant Fourier coefficient as $\mathbb{E}_k(\tau)$, thus one may apply Siegel's theorem.

Example 2. For $\mathfrak{a} \subset \mathcal{O}$ consider

$$\mathbb{E}_k(\tau, \mathfrak{a}) = \mathcal{N}(\mathfrak{a})^k \sum_{(\lambda,\mu) \in \mathfrak{a} \times \mathfrak{a}/\mathcal{O}^\times} \mathcal{N}(\lambda\tau + \mu)^{-k}, \quad \text{for even integer } k > 2,$$

which actually only depends on the class $[\mathfrak{a}] \in Cl_K$; here $\mathcal{N}(\lambda\tau + \mu) = \prod_{\sigma : K \hookrightarrow \mathbb{R}} (\lambda^\sigma \tau_\sigma + \mu^\sigma)$ with $\tau = ((\tau_\sigma)_\sigma)$ and $\mathcal{O}^\times$ acts diagonally on $\mathfrak{a} \times \mathfrak{a}$.

Then $\mathbb{E}_k(\tau) := \sum_{[\mathfrak{a}] \in Cl_K} \mathbb{E}_k(\tau, \mathfrak{a})$ restricted to $\mathbb{H}_{\mathbb{Q}}$ has q-expansion

$$(\Delta^* \mathbb{E}_k)(z) = \zeta_K(k) + \left(\frac{(2\pi i)^k}{(k-1)!} \right)^r d_K^{\frac{2k-1}{2}} \sum_{n=1}^{\infty} \sum_{\substack{0 \ll \xi \in \mathfrak{D}^{-1} \\ \operatorname{Tr}(\xi) = n}} \sigma_{k-1}(\xi \mathfrak{D}) q^n$$

where $\zeta_K = \sum_{\mathfrak{b} \subset \mathcal{O}} \mathcal{N}(\mathfrak{b})^{-s}$, $Re(s) > 1$, denotes the Dedekind zeta-function of K and $\sigma_k(\mathfrak{b}) := \sum_{\mathfrak{a} | \mathfrak{b}} \mathcal{N}(\mathfrak{a})^k$.

Thus we obtain Siegel's formula

$$\left(\frac{\Gamma(k)}{(2\pi i)^k}\right)^r \frac{\zeta_K(k)}{d_K^{\frac{2k-1}{2}}} = -\sum_{j=1}^{kd} \frac{c_{kd,j}}{c_{kd,0}} \sum_{\substack{0 \ll \xi \in \mathfrak{D}^{-1} \\ \mathrm{Tr}(\xi)=j}} \sigma_{k-1}(\xi\mathfrak{D}) \in \mathbb{Q}.$$

Remark 2. Siegel applies the same technique to (values of) partial ζ-functions, thereby showing the rationality of the L-values $L(1-k,\varepsilon)$ to be defined below.

Let I_0 (resp. $I_0(\mathfrak{f})$) denote the group of fractional ideals of K (prime to $\mathfrak{f}$). Also we write A_0 ($A_0(\mathfrak{f})$) for the submonoid of integral ideals in $\mathcal{O}$ (prime to $\mathfrak{f}$). Then $G_\mathfrak{f} := A_0(\mathfrak{f})/\sim_\mathfrak{f}$ is the strict ray class group, where $\mathfrak{a} \sim_\mathfrak{f} \mathfrak{b} \Longleftrightarrow \mathfrak{a}\mathfrak{b}^{-1} = (\alpha)$ for some $0 \ll \alpha \in 1 + \mathfrak{f}\mathfrak{b}^{-1}$.

For a function $\varepsilon : G_\mathfrak{f} \longrightarrow \mathbb{C}$ we set

$$L(s,\varepsilon) := \sum_{\mathfrak{a}\in A_0(\mathfrak{f})} \varepsilon(\mathfrak{a})\mathfrak{N}\mathfrak{a}^{-s} = \sum_{[\mathfrak{a}]=a\in G_\mathfrak{f}} \varepsilon(a)\zeta(\mathfrak{a},\mathfrak{f},s)$$

where

$$\zeta(\mathfrak{a},\mathfrak{f},s) = \sum_{\mathfrak{b}\in A_0(\mathfrak{f}),\, \mathfrak{b}\sim_\mathfrak{f}\mathfrak{a}} \mathcal{N}\mathfrak{b}^{-s}.$$

Then, setting $\mathfrak{b} := \mathfrak{a}\mathfrak{D}^{-1}\mathfrak{f}^{-1}$, for $k \geq 1$ Siegel constructs a Hilbert modular Eisenstein series $\mathbb{E}_k(\mathfrak{a},\mathfrak{f})(\tau)$ whose pullback $(\Delta^*\mathbb{E}_k(\mathfrak{a},\mathfrak{f}))(z)$ ($= \varphi(z)$ in [13, (98)]) is an elliptic modular form of weight kr

$$\left(\frac{\Gamma(k)}{(2\pi i)}^k\right)^r d_K^{\frac{2k-1}{2}} \mathcal{N}(\mathfrak{f})^{k-1}\mathcal{N}(\mathfrak{b})^k \left(\sum_{\substack{(x,\lambda)\in K^2/\mathcal{O}_+^\times \\ x\in\mathfrak{a}\mathfrak{D}^{-1},\lambda\in\mathfrak{b}}} e^{2\pi i \mathrm{Tr}\lambda} \mathcal{N}(xz+\lambda)^{-k}\mathcal{N}|xz+\lambda|^{-s} \right)_{|s=0},$$

where $\mathcal{O}_+^\times := \{\alpha \in \mathcal{O}^\times | \alpha = 1 \mod \mathfrak{f}, \alpha \gg 0\}$, and has Fourier coefficients (loc. cit., (100), (101) and the last displayed formula on page 19)

$$a_0 = \zeta(\mathfrak{a},\mathfrak{f},1-k)\ (k > 2, \text{ if } K = \mathbb{Q},\ \mathfrak{f} = 1),$$
$$a_n \in \mathbb{Q}.$$

Thus the rationality of the partial zeta values follows. Moreover, if $\varepsilon : G_\mathfrak{f} \to \mathbb{Q}$ takes values in the rational numbers, we obtain

Theorem I. $L(1-k,\varepsilon) \in \mathbb{Q}$ *for all* $k \geq 1$ $(K \neq \mathbb{Q})$.

Furthermore, we can extend the definition of $L(s,\varepsilon)$ to all functions $\varepsilon : G_\mathfrak{f} \to V$ with values in a $\mathbb{Q}$-vectorspace V, e.g. $\mathbb{Q}_p$, by setting

$$L(s,\varepsilon) := \sum_{a\in G_\mathfrak{f}} \zeta(\mathfrak{a},\mathfrak{f},s)\varepsilon(a) \in V.$$

2. Congruences

In a second step we are now going to discuss the type of congruences which Deligne–Ribet prove in order to show the existence of p-adic L-functions. To this end let

$$G = \varprojlim_n G_{\mathfrak{f}p^n}$$

be the strict ray class group of conductor $\mathfrak{f}p^\infty$, which by class field theory corresponds to the maximal abelian extension of K, which is unramified at all finite places v not dividing $\mathfrak{f}p$ (infinite places are allowed to ramify!). Let

$$\mathcal{N} : G \to \mathbb{Z}_p^\times$$

be the unique continuous character which is compatible with the usual norm map $\mathcal{N} : A_0(\mathfrak{f}p) \to \mathbb{Z}_p$ and the natural map $A_0(\mathfrak{f}p) \to G$. Note that $\mathcal{N}$ is actually nothing else than the p-cyclotomic character of K. For $\varepsilon : G_\mathfrak{f} \to \mathbb{Q}_p$ and $c \in G$ we set

$$\Delta_c(1-k,\varepsilon) := L(1-k,\varepsilon) - \mathcal{N}c^k L(1-k,\varepsilon_c) \in \mathbb{Q}_p$$

for $k \geq 1$, where $\varepsilon_c(g) := \varepsilon(\bar{c}g)$ if $\bar{c}$ denotes the image of c under the canonical projection $G \twoheadrightarrow G_f$. Note that for a character ε this is equal to

$$(1 - \mathcal{N}c^k\, \varepsilon(\bar{c}))\, L(1-k,\, \varepsilon).$$

For a family $(\varepsilon_k : G_\mathfrak{f} \to \mathbb{Q}_p)_{k\geq 1}$ (almost all zero) we define a map φ on $A_0(\mathfrak{f})$ via

$$\varphi(\mathfrak{a}) = \sum_{k\geq 1} \varepsilon_k(\mathfrak{a}) \mathcal{N}\mathfrak{a}^{k-1} \in \mathbb{Q}_p.$$

Theorem II. *If $\varphi(\mathfrak{a}) \in \mathbb{Z}_p$ for all $\mathfrak{a} \in A_0(\mathfrak{f})$, then*

$$\Delta := \sum_{k\geq 1} \Delta_c(1-k,\ \varepsilon_k) \in \mathbb{Z}_p \ \textit{for all}\ \ c \in G.$$

These kinds of congruences were (axiomatically) first introduced by Coates [1]. In the case $K = \mathbb{Q}$ they are nothing else than the "generalized Kummer congruences" as considered by Mazur and others and they can be rather easily proved from the fact that over $\mathbb{Q}$ the values $L(1-k,\varepsilon)$ for $k \geq 1$ can be expressed in terms of Bernoulli polynomials; see [2, §1]. Thus, from now on we shall concentrate on the case $K \neq \mathbb{Q}$.

For the purpose of p-adic L-functions we now assume that all primes lying above p divide the conductor $\mathfrak{f} \subset \mathcal{O}$. Note that a locally constant function $\varepsilon : G \to \mathbb{Q}_p$ factorizes over $G_{\mathfrak{f}'}$ for some $\mathfrak{f}|\mathfrak{f}'|(\mathfrak{f}p)^n$, $n \geq 1$, which allows us to define $\Delta_c(1-k,\varepsilon)$ (independently of the choice of $\mathfrak{f}'$, as $A_0(\mathfrak{f}') = A_0(\mathfrak{f})$ for any such $\mathfrak{f}'$). Thus we obtain a distribution for any $k \geq 1$, $c \in G$,

$$\begin{aligned}\mu_{c,k} : \{\varepsilon : G \to \mathbb{Q}_p \text{ locally constant}\} &\longrightarrow \mathbb{Q}_p \\ \varepsilon &\longmapsto \Delta_c(1-k,\varepsilon).\end{aligned}$$

Setting $\varepsilon_k := \varepsilon : G \to \mathbb{Z}_p$ (locally constant) and $\varepsilon_i := 0$ for $i \neq k$, it follows from Theorem II that $\mu_{c,k}(\varepsilon) \in \mathbb{Z}_p$, i.e., the distribution $\mu_{c,k}$ is bounded or in other words:

Theorem III. *$\mu_{c,k}$ is a measure on G with values in $\mathbb{Z}_p$. Moreover, $\mu_{c,k} = \mathcal{N}^{k-1} \cdot \mu_{c,1}$.*

Proof. It suffices to show for $k \geq 1$ and $n \geq 1$ that

$$\int \varepsilon \mathcal{N}^{k-1} d\mu_{c,1} \equiv \Delta_c(1-k, \varepsilon) \bmod p^n \, \mathbb{Z}_p$$

for each locally constant $\varepsilon : G \to \mathbb{Z}_p$. For this, let $\eta : G \to \mathbb{Z}_p$ be a locally constant function such that

$$\eta \equiv \mathcal{N}^{k-1} \bmod p^n.$$

Taking $\varepsilon_1 := \varepsilon_\eta, \varepsilon_k := -\varepsilon$, and $\varepsilon_i := 0$ for $i \neq 1, k$, the function $\varphi = \varepsilon\eta - \varepsilon\mathcal{N}^{k-1}$ takes values in $p^n\mathbb{Z}_p$, and Theorem II implies:

$$\Delta_c(0, \varepsilon\eta) - \Delta_c(1-k, \varepsilon) \in p^n\mathbb{Z}_p,$$

which gives

$$\Delta_c(1-k, \varepsilon) \equiv \Delta_c(0, \varepsilon\eta) = \int \varepsilon\eta d\mu_{c,1} \equiv \int \varepsilon \mathcal{N}^{k-1} d\mu_{c,1} \bmod p^n\mathbb{Z}_p. \qquad \square$$

Setting $\lambda_c := \mathcal{N}^{-1}\mu_{c,1} \in \mathbb{Z}_p[[G]]$ for $c \in G$ we obtain

$$\int \mathcal{N}^k \varepsilon d\lambda_c = \Delta_c(1-k, \varepsilon).$$

Then $\lambda := \frac{1}{1-c}\lambda_c$ in the total quotient ring of $\mathbb{Z}_p[[G]]$ is a *pseudo-measure* à la Serre [11] for $1-c$ a non-zero divisor in $\mathbb{Z}_p[[G]]$:

$$(1-c')\lambda = \lambda_{c'} \in \mathbb{Z}_p[[G]] \text{ for all } c' \in G.$$

Also, it can be deduced that for $k \geq 1$

$$\int \mathcal{N}^k \varepsilon d\lambda = L(1-k, \varepsilon).$$

Indeed, if ε belongs to the set of continuous characters X_G of G (with values in $\mathbb{C}_p^\times$) we calculate

$$\int \mathcal{N}^k \varepsilon d\lambda = \frac{\int \mathcal{N}^k \varepsilon d\lambda_c}{1 - \mathcal{N}c^k\varepsilon(c)} = \frac{L(1-k, \varepsilon) - \mathcal{N}c^k L(1-k, \varepsilon_c)}{1 - \mathcal{N}c^k\varepsilon(c)} = L(1-k, \varepsilon),$$

where we choose a c which does not lie in the kernel of $\mathcal{N}^k\varepsilon$.

Setting

$$\mathcal{L} = \mathcal{L}_{\mathfrak{f}} : X_G \backslash \{\mathbb{1}\} \longrightarrow \mathbb{C}_p^\times$$

$$\varphi \longmapsto \int \varphi d\lambda$$

we get the p-adic zeta-function of K as

$$\zeta_{K,p}(s) = \mathcal{L}(\langle\chi_K\rangle^{1-s}), \quad s \in \mathbb{Z}_p\backslash\{1\},$$

satisfying

$$\zeta_{K,p}(1-k) = \mathcal{L}(\omega_K^{-k}\chi_K^k) = L_{\mathfrak{f}}(1-k, \omega_K^{-k})$$
$$= \zeta_{K,\mathfrak{f}}(1-k) \quad \text{if } k|\#\omega_K.$$

Here χ_K denotes the p-cyclotomic character, ω_K the Teichmüller character of K and $\langle\,\rangle$ the projection from $\mathbb{Z}_p^\times$ to $1 + p\mathbb{Z}_p$. More generally, the p-adic L-function for $\phi \in X_G$ arises as

$$L_{p,K}(s, \phi) := \mathcal{L}(\phi\langle\chi_K\rangle^{1-s})$$

with

$$L_{p,K}(1-k, \phi) = L_{\mathfrak{f}}(1-k, \phi\omega_K^{-k}),$$

where $L_{\mathfrak{f}}(s, \phi\omega_K^{-k})$ is the Artin L-function attached to $\phi\omega_K^{-k}$ without the Euler factors of the places dividing $\mathfrak{f}$.

Consider the following decomposition and projection

$$\alpha : G \cong A \times \Gamma \twoheadrightarrow \Gamma \xrightarrow{\sim} \mathbb{Z}_p,$$

$\alpha(\gamma) = 1$ for a fixed topological generator γ of Γ, where $A = ker(\langle\,\rangle)$ where we simply write $\langle\,\rangle$ also for the composite $\langle\,\rangle \circ \mathcal{N}$. For the convenience of the reader we recall the following well-known

Proposition 2.1 ([10, (4.8)–(4.10)]). *Let $\varepsilon : G \to \mathbb{Z}_p[\varepsilon]^\times$ be an (even) character of finite order.*

(i) *Then, for each c, there exist unique $H'_{\varepsilon,c}, G'_{\varepsilon,c} \in \mathbb{Z}_p[\varepsilon][[X]]$ such that*

$$\int \langle x\rangle^{1-s}\varepsilon(x)d\lambda_c(x) = G'_{\varepsilon,c}(\langle\gamma\rangle^{1-s} - 1),$$
$$1 - \varepsilon(c)\langle c\rangle^{1-s} = H'_{\varepsilon,c}(\langle\gamma\rangle^{1-s} - 1).$$

(ii) *If $\varepsilon_{|A} \not\equiv \mathbb{1}$, then*

$$L_{p,K}(s, \varepsilon) = F_\varepsilon(\langle\gamma\rangle^{1-s} - 1)$$

for some $F_\varepsilon \in \mathbb{Z}_p[\varepsilon][[X]]$. Clearly, $F_\varepsilon = \frac{G'_{\varepsilon,c}}{H'_{\varepsilon,c}}$ for all c such that $H'_{\varepsilon,c} \neq 0$ even though $H'_{\varepsilon,c}$ need not be a unit.

(iii) *If $\varepsilon_{|A} \not\equiv \mathbb{1}$ and $\theta : G \to \mathbb{C}_p^\times$ is a character of finite order such that $\theta_{|A} = \mathbb{1}$ is trivial, then*

$$F_{\theta\varepsilon}(X) = F_\varepsilon(\zeta(1+X) - 1) \quad \text{with} \quad \zeta := \theta(\gamma) \in \mu_{p^\infty}.$$

Remark 3. Setting

$$H_\epsilon = \begin{cases} -H'_{\epsilon,\gamma}, & \text{if } \varepsilon_{|A} \equiv 1\!\!1, \\ 1, & \text{otherwise,} \end{cases}$$

and

$$G_\epsilon = \begin{cases} -G_{\epsilon,\gamma}, & \text{if } \varepsilon_{|A} \equiv 1\!\!1, \\ F_\epsilon, & \text{otherwise,} \end{cases}$$

we obtain the unique power series in [14] such that

$$L_{p,K}(s,\varepsilon) = \frac{G_\varepsilon(\langle\gamma\rangle^{1-s} - 1)}{H_\varepsilon(\langle\gamma\rangle^{1-s} - 1)}.$$

Proof (sketch). Note that for any $\psi \in X_G$ there exists a character ψ_A which is trivial on Γ and such that

$$\psi(g) = \psi_A(g)\ \psi(\gamma)^{\alpha(g)}$$

for all $g \in G$.

We will use the following well-known

Fact. Let μ be a measure on G. Then

$$\int \psi d\mu = F(\psi(\gamma) - 1)$$

where $F = F_{\psi_A,\mu} := \sum_{n\geq 0} a_n X^n \in \mathbb{Z}_p[\psi][[X]]$ with

$$a_n = \int_G \psi_A(x)\binom{\alpha(x)}{n} d\mu(x) \in \mathbb{Z}_p[\psi].$$

Taking $\psi = \langle\,\rangle^{1-s}$ and $\mu = \varepsilon\lambda_c$ we have

$$\int \langle\,\rangle^{1-s}\varepsilon d\lambda_c = F(\langle\gamma\rangle^{1-s} - 1).$$

Setting $G'_{\varepsilon,c} := F = F_{\langle\,\rangle^{1-s}{}_A,\varepsilon\lambda_c}$ and $H'_{\varepsilon,c} := 1 - \varepsilon(c)(X+1)^\alpha$, if $\langle c\rangle = \langle\gamma\rangle^\alpha$ for $\alpha \in \mathbb{Z}_p$, we obtain (i).

(ii) By (4.3) in (loc. cit.) there exists a measure (not just pseudo-measure) μ' on G such that

$$L_{p,K}(s,\epsilon) = \int \langle\,\rangle^{1-s}\varepsilon d\mu'$$

for all ϵ which are non-trivial on A. Now apply the above fact to the measure $\mu = \varepsilon\mu'$ and $\psi = \langle\,\rangle^{1-s}$.

(iii) By definition we have

$$F_{\theta\varepsilon}(\langle\gamma\rangle^{1-s} - 1) = \int \langle x\rangle^{1-s}\theta\varepsilon d\mu'$$

$$= F_{(\langle\,\rangle^{1-s}\theta)_A,\varepsilon\mu'}(\langle\gamma\rangle^{1-s}\theta(\gamma) - 1)$$

by the above fact. But

$$F_\varepsilon = F_{\langle\,\rangle_A^{1-s},\varepsilon\mu'} = F_{(\langle\,\rangle^{1-s}\theta)_A,\varepsilon\mu'}$$

because $\theta_A = \mathbb{I}$ and the claim follows. □

For the proof of Theorem II we need a q-expansion principle which we shall state now, but explain in more detail later.

For simplicity we assume $\mathfrak{f} = N$, a positive integer, i.e.,

$$\Gamma_{00}(N) = \left\{ \begin{pmatrix} a & b \\ c & d \end{pmatrix} \in SL_2(K) | \begin{matrix} a, d \in 1 + N\mathcal{O} \\ b \in \mathfrak{D}^{-1}, c \in N\mathfrak{D} \end{matrix} \right\}$$

Let $F : \mathbb{H}_K \to \mathbb{C}$ be a Hilbert modular form of weight k with Fourier expansion

$$F_1 = a_0 + \sum_{\mu\in\mathcal{O},\ \mu\gg 0} a_\mu, \quad q^\mu = e^{2\pi i Tr(\mu\tau)}$$

at 1 and

$$F_\alpha$$

at other cusps α. Actually, the cusps we are interested in are all induced from finite ideles $\alpha \in \hat{K}^\times$ [2, (5.3)], which in turn induce elements $c(\alpha)$ in G (called $j(\alpha)$ just below [2, (2.22)]).

Perhaps the hardest part in [2] is the proof of the geometric irreducibility of the fibres of moduli space of Hilbert-Blumenthal Abelian Varieties (HBAV), which implies the following

q-expansion-principle (Rapoport, Deligne–Ribet)

I. If the q-expansion coefficients of F at one cusp are all rational, then this is true for all cusps.

   Now let F_k, $k \geq 0$, be forms of weight k, almost all zero, such that $F_{k,1} \in$ "$\mathbb{Q}[[q]]$" and set

$$S(\alpha) := \sum_{k\geq 0} \mathcal{N}\alpha_p^{-k} F_{k,\alpha} \in \mathbb{Q}_p[[q]],$$

   which corresponds (only) to a p-adic modular form, as it mixes up different weights! Here α_p denotes the p-component of $\alpha \in \hat{K}^\times$.

II. If $S(\alpha)$ has coefficients in $\mathbb{Z}_p$ for one α, then the same holds for all α.

III. Assume that all non-constant coefficients of $S(\alpha)$ lie in $\mathbb{Z}_p$. Then this holds also for $S(\beta)$ at any other cusp β and

$$a_0(S(\alpha)) - a_0(S(\beta) \in \mathbb{Z}_p.$$

Now the idea of the proof of Theorem II consists of finding suitable Hilbert modular forms F_{k,ε_k} such that for all cusps α their Fourier expansions are essentially of the form

$$(F_{k,\varepsilon_k})_\alpha = \mathcal{N}c(\alpha)^k \mathcal{N}\alpha_p{}^k \left\{ L(1-k, \varepsilon_{k,c(\alpha)}) + \sum_{\mu \gg 0} \left(\sum_{?} \varepsilon_{k,c(\alpha)} \mathcal{N}^{k-1}(\mu ?) \right) q^\mu \right\},$$

for $k \geq 1$ ($F_0 = 0$). Then, by the assumption of Theorem II all non-constant coefficients belong to $\mathbb{Z}_p$. Thus the q-expansion principle III implies that

$$\begin{aligned}\Delta = \sum_{k \geq 1} \Delta_{c(\alpha)}(1-k, \varepsilon_k) &= \sum_{k \geq 1} (L(1-k, \varepsilon_k) - \mathcal{N}c(\alpha)^k L(1-k, \varepsilon_{k,c(\alpha)}) \\ &= a_0(S(1)) - a_0(S(\alpha)) \in \mathbb{Z}_p\end{aligned}$$

and Theorem II follows. □

The needed Eisenstein series F_{k,ε_k} for arbitrary locally constant functions ε (these are needed for the distribution property) are similar to those studied by Siegel (see page 5), but while for rationality it sufficed for Siegel to realise the special L-values as constant terms of elliptic modular forms, for the purpose of congruences Deligne–Ribet had to find Hilbert modular forms with the corresponding L-values as constant terms; they are constructed in [2, §6] in an adelic setting! Also we should note that for $k = 1$ such $F_{1,\varepsilon}$ apparently does not exist and Deligne and Ribet have to circumvent this problem, which leads to some technical problems which we do not want to discuss here.

The rest of this survey is devoted to explain the q-expansion principle in a little more detail.

2.1 *The moduli space*

References for this paragraph concerning the moduli problem of classifying abelian varieties with real multiplication are [9], [8, §1], [2, §4].

Recall that a Hilbert-Blumenthal abelian variety (HBAV) relative to $\mathcal{O}$ over a base S is an abelian scheme X/S, furnished with a homomorphism $m : \mathcal{O} \hookrightarrow \mathrm{End}(X)$ making its Lie algebra $Lie(X/S)$ into a locally free $\mathcal{O} \otimes \mathcal{O}_S$-module of rank 1, in particular the relative dimension of X over S equals $r = [K : \mathbb{Q}]$. Fixing a polarization module $\mathfrak{c} \subset K$, and considering for simplicity only the case $S = \mathrm{Spec}(R)$ for any ring R, a $\mathfrak{c}$-polarization of X/S is a (positive) isomorphism

$$\lambda : \mathcal{P}(X) := \mathrm{Hom}_{\mathcal{O}}(X, X^t) \xrightarrow{\sim} \mathfrak{c}$$

of $\mathcal{O}$-modules, where X^t denotes the dual abelian variety of X.

Finally, for $N \geq 1$ a $\Gamma_{00}(N)$-structure of X is an $\mathcal{O}$-linear immersion $\mathcal{O} \otimes \mu_{N/S} \hookrightarrow X$. Now consider the moduli problem, given by the functor:

$$F(\mathfrak{c}, N) : \text{Schemes} \longrightarrow \text{Sets}$$

$$S \longmapsto \left\{ \begin{array}{l} \text{isomorphism classes of triples} \\ (X, \lambda, i) \text{ with } X \text{ a HBAV } /S, \\ \lambda \text{ a } \mathfrak{c}-\text{polarization}, \\ i \text{ a } \Gamma_{00}(N)\text{-structure} \end{array} \right\}.$$

The moduli stack $\mathfrak{M}(\mathfrak{c}, N)$ defined by this functor is an algebraic stack, smooth of relative dimension r over $\mathbb{Z}$ and is *represented* by an algebraic space, also denoted $\mathfrak{M}(\mathfrak{c}, N)$, smooth of relative dimension r over $\mathbb{Z}$ for $N \geq 4$. In particular, there is a unique universal $\mathfrak{c}$-polarized HBAV with $\Gamma_{00}(N)$-structure

$$\begin{array}{ccc} \varphi^*\mathcal{X} & \longrightarrow & \mathcal{X} := (X_{\text{univ}}, \lambda_{\text{univ}}, i_{\text{univ}}) \\ \downarrow & & \downarrow \pi \\ S & \xrightarrow{\varphi} & \mathfrak{M}(\mathfrak{c}, N) \end{array}$$

and all elements in $F(\mathfrak{c}, N)(S)$ arise as pullback $\varphi^*\mathcal{X}$ for some $\varphi \in \text{Mor}(S, \mathfrak{M}(\mathfrak{c}, N))$. In particular, for a field L, the L-valued points

$$\mathfrak{M}(\mathfrak{c}, N)(L)$$

consists of $\mathfrak{c}$-polarized HBAV with $\Gamma_{00}(N)$-structure defined over L.

2.2 *Hilbert-modular forms (HMF)*

Assume $N \geq 4$ and denote by $\underline{\omega}$ the sheaf $\pi_*(\Omega^1_{X_{\text{univ}}/\mathfrak{M}})$ on $\mathfrak{M} := \mathfrak{M}(\mathfrak{c}, N)$ and let

$$\underline{\omega}(k) := \bigotimes_\sigma \underline{\omega}(\sigma)^k$$

be $\underline{\omega}$ extended by the character $\mathcal{N}^k$, which is an invertible sheaf over $\mathfrak{M}_R := \mathfrak{M} \otimes R$, on which $(\mathcal{O} \otimes R)^\times$ acts by $\mathcal{N}^k$, see [5, §1.3] for details. Here we fix a ground ring R and for every R-algebra R' and as before $\mathcal{N}$ denotes the norm character $\mathcal{O} \otimes R' \to R'$. Then the $\mathfrak{c}$-HMFs of weight k on $\Gamma_{00}(N)$ defined over R are by definition the global sections

$$\mathcal{M}_k(\mathfrak{c}, N, R) := H^0(\mathfrak{M}(\mathfrak{c}, N)_R, \underline{\omega}(k)).$$

This can also be interpreted (and this makes even sense for $N \geq 1$) as a rule F which assigns to a quadruple

$$(X, \lambda, \omega, i)$$

over R' an element $F(X, \lambda, \omega, i) \in R'$, where

1. (X, λ) is a $\mathfrak{c}$-polarized HBAV over an R-algebra R',
2. ω is a nowhere vanishing differential on X and
3. i is a $\Gamma_{00}(N)$-structure

subject to the following conditions

(i) $F(X, \lambda, \omega, i)$ depends only on the R'-isomorphism-class of (X, λ, ω, i),
(ii) Formation of $F(X, \lambda, \omega, i)$ is compatible with any extension of scalars $R'' \to R'$ of R-algebras,
(iii) $F(X, \lambda, a^{-1}\omega, i) = \mathcal{N}(a)^k F(X, \lambda, \omega, i)$ for all $a \in (\mathcal{O} \otimes R')^\times$.

2.3 *q-expansions*

Let α be in $\hat{K}^\times$, $\mathfrak{a} = (\alpha)$ the (fractional) ideal generated by it and $\mathfrak{b} := \mathfrak{a}\mathfrak{c}^{-1}$. Then there is an HBAV-analogue of the Tate-elliptic curve: more precisely, there exist a certain Laurent series ring $\mathbb{Z}((q)) := \mathbb{Z}((q^\mu, \mu \in \mathfrak{ab}))$ and a HBAV $\mathrm{Tate}_\alpha(q)$ defined over $\mathbb{Z}((q))$ together with a canonical polarization λ_{can}, a differential $\omega_{\mathfrak{a}}$ and a Γ_{00}-structure i_α. Then the value

$$F(\mathrm{Tate}_\alpha(q), \lambda_{\mathrm{can}}, \omega_\alpha, i_\alpha) \in R \otimes \mathbb{Z}((q)) \subseteq R((q))$$

has by definition of the latter ring a q-expansion

$$F_\alpha = \sum_{\mu \in \mathfrak{ab}} a_\mu(F_{\mathfrak{a}}) q^\mu$$

at the "cusp α" (actually there are some choices involved which for simplicity we do not want to mention here).
In the case $K \neq \mathbb{Q}$, one knows that

$$a_\mu(F_\alpha) = 0 \text{ unless } \mu = 0 \text{ or } \mu \gg 0.$$

Theorem IV (Ribet). *The geometric fibres of $\mathfrak{M}(\mathfrak{c}, N)$ over $Spec(\mathbb{Z})$ are all geometrically irreducible.*

From the standard geometric argument[1] in [9], see also [6, thm. 1.6.1] or [7] one obtains the following

Corollary (q-expansion principle I). *Fix $N \geq 1$, $k \geq 0$ and R.*

(i) *Then for all cusps α as above the map*

$$\begin{aligned} \mathcal{M}_k(\mathfrak{c}, N, R) &\hookrightarrow R((q)) \\ F &\longmapsto F_\alpha \end{aligned}$$

is injective.

[1] Roughly speaking the idea is as follows: One has to show that the morphism Spec $(\mathbb{Z}((q))) \to \mathfrak{M}(\mathfrak{c}, N)$ corresponding to the point $\mathrm{Tate}_\alpha(q)$ of $\mathfrak{M}$ has the following topological property: its image contains the generic point (of each fibre). Then the vanishing of the modular form F in $\mathrm{Tate}_\alpha(q)$ implies the vanishing of F in each of these generic points. By the smoothness and irreducibility (which imply the absence of embedded components) this implies the vanishing of F everywhere.

(ii) *Let $R \subset R'$ and let F be in $\mathcal{M}_k(\mathfrak{c}, N, R')$. If the coefficients of F_α are all in R, then F arises actually from a unique HMF defined already over R, i.e., the sequence*

$$0 \to \mathcal{M}_k(\mathfrak{c}, N, R) \to \mathcal{M}_k(\mathfrak{c}, N, R') \to R'((q))/R((q))$$

is exact.

Remark 4. For $K \neq \mathbb{Q}$ the modular forms $\mathcal{M}_k(\mathfrak{c}, N, \mathbb{C})$ over $\mathbb{C}$ can also be identified with the set of holomorphic functions $F : \mathbb{H}_K \to \mathbb{C}$ such that

$$F_{|M}(\tau) = \mathcal{N}(\gamma\tau + \delta)^{-k} F\left(\frac{\alpha\tau + \beta}{\gamma\tau + \delta}\right) = F(\tau)$$

for all

$$M = \begin{pmatrix} \alpha & \beta \\ \gamma & \delta \end{pmatrix} \in \Gamma_{00}(N).$$

To this end one defines

$$F(\tau) := F\left(\frac{K \otimes \mathbb{C}}{2\pi i(\mathfrak{a}^{-1}\mathfrak{D}^{-1} + \mathfrak{a}\mathfrak{c}^{-1}\tau)}, \lambda_{\mathrm{can}}, \omega_{\mathrm{can}}, i_{\mathrm{can}}\right)$$

for $\omega_{\mathrm{can}} = dz$ and certain canonical choices $\lambda_{\mathrm{can}}, i_{\mathrm{can}}$ (cf. [2] before (5.7), compare also with [4, §4.13]).

By the translation-invariance with respect to elements of $(\mathfrak{a}\mathfrak{b}\mathfrak{D})^{-1}$ thus F admits a Fourier-expansion

$$F_\alpha(\tau) = \sum_{\substack{\mu \in \mathfrak{a}\mathfrak{b} \\ \mu = 0 \\ \text{or} \\ \mu \gg 0}} c_\mu(F_\alpha) q^\mu \quad (q^\mu := e^{2\pi i\, tr(\tau\cdot\mu)}) \text{ at the cusp } \alpha.$$

By a GAGA-like result one has

$$c_\mu(F_\alpha) = a_\mu(F_\alpha),$$

where a_μ is the Fourier-coefficient defined earlier.

2.4 *p-adic modular forms*

Let R be a p-adic ring, i.e.,

$$R = \varprojlim_n R/p^n R.$$

Then the R-module $V(N, R)$ of p-adic modular forms on $\Gamma_{00}(N)$ over R (in the sense of Katz) consists of functions f on isomorphism classes of triples (X, λ, i) such that (X, λ) is a $\mathfrak{c}$-polarized HBAV over a p-adic R-algebra R' and i is a $\Gamma_{00}(Np^\infty)$-structure, i.e., a compatible system of $\Gamma_{00}(Np^n)$-structures, $n \geq 0$, subject to the condition that f is compatible with scalar extensions $R' \to R''$ of p-adic R-algebras.

Note first that there is no concept of weights anymore. Secondly, if p is nilpotent in R, any R-algebra R' is automatically p-adic, thus there is a natural identification

$$\mathcal{M}_0(\mathfrak{c}, Np^\infty, R/p^nR) \cong V(N, R/p^nR)$$

of weight 0 $\mathfrak{c}$-HMF with p-adic modular forms, both over R/p^nR. Thus we obtain the following interpretation for $N \geq 4$. Let $\mathfrak{M}_R^{p\text{-adic}}$ be the formal scheme $\{\mathfrak{M}(\mathfrak{c}, Np^\infty)_{R/p^nR}\}_{n\in\mathbb{N}}$ with structure sheaf $\mathcal{O}_{\mathfrak{M}^{p\text{-adic}}}$. Then the global sections

$$\begin{aligned} H^0(\mathfrak{M}_R^{p\text{-adic}}, \mathcal{O}_{\mathfrak{M}^{p\text{-adic}}}) &\cong \varprojlim_n H^0(\mathfrak{M}(\mathfrak{c}, Np^\infty)_{R/p^nR}, \mathcal{O}_{\mathfrak{M}^{p\text{-adic}}}) \\ &\cong \varprojlim_n \mathcal{M}_0(\mathfrak{c}, Np^\infty, R/p^nR) \\ &\cong \varprojlim_n V(N, R/p^n) \\ &\cong V(N, R) \end{aligned}$$

are nothing else than the p-adic modular forms over R.

Now we concentrate on the case $R = \mathbb{Z}_p$. To a finite idele $\alpha \in \hat{K}^\times$ one can also attach a *p-adic cusp* [2, (5.11)] and we have the q-expansion-maps

$$\begin{aligned} \phi_\alpha : V(N, \mathbb{Z}_p) &\longrightarrow \mathbb{Z}_p[[q]] \\ f &\longmapsto f(\mathrm{Tate}_\alpha(q), \lambda_{\mathrm{can}}, i_\alpha) \end{aligned}$$

and $\phi_\alpha \otimes \mathbb{Q}_p : V(N, \mathbb{Q}_p) := V(N, \mathbb{Z}_p) \otimes_{\mathbb{Z}_p} \mathbb{Q}_p \longrightarrow \mathbb{Q}_p \otimes_{\mathbb{Z}_p} \mathbb{Z}_p[[q]] \subseteq \mathbb{Q}_p[[q]]$.

Theorem 2.4.

(i) *ϕ_α is injective.*
(ii) *coker (ϕ_α) does not have any p-torsion.*
(iii) *The sequence*

$$0 \to V(N, \mathbb{Z}_p) \to V(N, \mathbb{Q}_p) \stackrel{\phi_\alpha \otimes \mathbb{Q}_p}{\longrightarrow} \mathbb{Q}_p[[q]]/\mathbb{Z}_p[[q]]$$

is exact, i.e., a p-adic modular form $f \in V(N, \mathbb{Q}_p)$ all of whose q-expansion coefficients at one p-adic cusp α belong to $\mathbb{Z}_p$ is already defined over $\mathbb{Z}_p$.

Sketch of proof. Since ϕ_α is the inverse limit of the injective maps (by the Corollary of Ribet's theorem)

$$V(N, \mathbb{Z}_p/p^n\mathbb{Z}_p) = \mathcal{M}_0(\mathfrak{c}, Np^\infty, \mathbb{Z}_p/p^n) \hookrightarrow \mathbb{Z}_p/p^n\mathbb{Z}_p[[q]]$$

it is clearly injective itself. Furthermore with $\mathfrak{M}(\mathfrak{c}, Np^\infty)/\mathbb{Z}$ also $\mathfrak{M}_{\mathbb{Z}_p}^{p\text{-adic}}/\mathbb{Z}_p$ is flat, because flatness is stable under base change, here by $(\mathbb{Z} \to)\mathbb{Z}_{(p)} \to \mathbb{Z}_p$. Hence we have the exact sequence

$$0 \to \mathcal{O}_{\mathfrak{M}^{p\text{-adic}}} \stackrel{p}{\to} \mathcal{O}_{\mathfrak{M}^{p\text{-adic}}} \to \mathcal{O}_{\mathfrak{M}^{p\text{-adic}}}/p \to 0.$$

Taking global sections gives

$$0 \to V(N, \mathbb{Z}_p) \overset{p}{\to} V(N, \mathbb{Z}_p) \to V(N, \mathbb{Z}_p/p\mathbb{Z}_p). \tag{1}$$

Now let $f(q) \in \mathbb{Z}_p[[q]]$ be a power series in q such that $pf(q) = g(q)$ for some $g \in V(N, \mathbb{Z}_p)$. Then the image of g in $V(N, \mathbb{Z}_p/p\mathbb{Z}_p)$ is zero by (i). From (1) we see that there is a unique $f \in V(N, \mathbb{Z}_p)$ such that $g = pf$ and with q-expansion $f(q)$. The claim (ii) follows. Finally, tensoring the exact sequence

$$0 \to V(\mathbb{Z}_p) \overset{\phi_\alpha}{\to} \mathbb{Z}_p[[q]] \to \text{coker } \phi_\alpha \to 0$$

with $\mathbb{Q}_p/\mathbb{Z}_p$ claim (iii) follows from (ii). □

Finally we remark that there is a canonical map

$$\mathcal{M}_k(\mathfrak{c}, N, R) \overset{\pi_k}{\to} V(N, R)$$

for any $k \geq 0$, induced by sending (X, λ, i) to $(X, \lambda, \omega(i), i)$ where $\omega(i)$ is a certain differential attached to i. Now we are able to prove the q-expansion principle III. In [2, (5.15)] it is explained that

$$S(\alpha) = \sum_{k \geq 1} \mathcal{N}\alpha_p{}^{-k} F_{k,\alpha} = \left(\sum_{k \geq 1} \pi_k F_k \right)_\alpha ,$$

where the last q-expansion is with respect to the p-adic cusp attached to α, while $F_{k,\alpha}$ denotes the expansion at the usual cusp. Setting F for $\sum_{k \geq 1} \pi_k F_k$, we have by the assumption of III that the q-expansion at the p-adic cusp α of the p-adic modular form $(F - t)$ has coefficients in $\mathbb{Z}_p$:

$$(F - t)_\alpha \in \mathbb{Z}_p[[q]],$$

where $t = a_0(F_\alpha)$ denotes the constant term at α, considered as weight zero HMF or constant p-adic modular form, respectively. Thus $F - t$ belongs to $V(N, \mathbb{Z}_p)$ by (iii) of the last theorem. Hence the q-expansion of $F - t$ at any other cusp β, say, also has coefficients in $\mathbb{Z}_p$. In particular, its constant term

$$\begin{aligned} a_0((F - t)_\beta) &= a_0(F_\beta) - a_0(t_\beta) \\ &= a_0(F_\beta) - t \\ &= a_0(F_\beta) - a_0(F_\alpha) \end{aligned}$$

belongs to $\mathbb{Z}_p$.

Acknowledgement

I would like to thank the organizers for the very instructive workshop in such a nice atmosphere and surroundings. Furthermore I am grateful to Thansis Bouganis and Ulrich Görtz for some discussions as well as to the referee for improving the presentation.

References

[1] J. Coates, p-adic L-functions and Iwasawa's theory, Algebraic number fields: L-functions and Galois properties (Proc. Sympos., Univ. Durham, Durham, 1975), Academic Press, London (1977) 269–353.

[2] P. Deligne and K. A. Ribet, Values of abelian L-functions at negative integers over totally real fields, *Invent. Math.*, **59** (1980) no. 3, 227–286.

[3] H. Hida, Elementary theory of L-functions and Eisenstein series, London Mathematical Society Student Texts, vol. 26, Cambridge University Press, Cambridge (1993).

[4] H. Hida, p-adic automorphic forms on Shimura varieties, Springer Monographs in Mathematics, Springer-Verlag (2004).

[5] H. Hida and J. Tilouine, Anti-cyclotomic Katz p-adic L-functions and congruence modules, *Ann. Sci. École Norm. Sup. (4)*, **26** (1993) no. 2, 189–259.

[6] N. Katz, p-adic properties of modular schemes and modular forms, Modular functions of one variable, III (Proc. Internat. Summer School, Univ. Antwerp, 1972) (Willem Kuyk and J.-P. Serre, eds.), Springer, Berlin, 1973, pp. 69–190. Lecture Notes in Math., Vol. 350.

[7] N. Katz, p-adic L-functions via moduli of elliptic curves, Algebraic geometry (*Proc. Sympos. Pure Math.*, Vol. 29, Humboldt State Univ., Arcata, Calif., 1974), *Amer. Math. Soc.*, Providence, R. I. (1975) 479–506.

[8] N. Katz, p-adic L-functions for CM fields, *Invent. Math.*, **49** (1978) no. 3, 199–297.

[9] M. Rapoport, Compactifications de l'espace de modules de Hilbert-Blumenthal, *Compositio Math.*, **36** (1978) no. 3, 255–335.

[10] K. A. Ribet, Report on p-adic L-functions over totally real fields, Journées Arithmétiques de Luminy (Colloq. Internat. CNRS, Centre Univ. Luminy, Luminy, 1978), Astérisque, 61, Soc. Math. France, Paris (1979) 177–192.

[11] J.-P. Serre, Sur le résidu de la fonction zêta p-adique d'un corps de nombres, *C. R. Acad. Sci. Paris Sér. A-B*, **287** (1978) no. 4, A183–A188.

[12] C. L. Siegel, Berechnung von Zetafunktionen an ganzzahligen Stellen, *Nachr. Akad. Wiss. Göttingen Math.-Phys. Kl. II*, (1969) 87–102.

[13] C. L. Siegel, Über die Fourierschen Koeffizienten von Modulformen, *Nachr. Akad. Wiss. Göttingen Math.-Phys. Kl. II*, (1970) 15–56.

[14] A. Wiles, The Iwasawa conjecture for totally real fields, *Ann. of Math. (2)*, **131** (1990) no. 3, 493–540.

Guwahati Workshop on Iwasawa Theory of Totally Real Fields No. 12, 2010, pp. 113–140.

The Main Conjecture

J. Coates[1] and R. Sujatha[2]

[1]Emmanuel College, Cambridge, CB2 3AP, England
[2]School of Mathematics, Tata Institute of Fundamental Research, Homi Bhabha Road, Colaba, Bombay 400 005, India
e-mail: J.H.Coates@dpmms.cam.ac.uk; sujatha@math.tifr.res.in

1. Introduction

In his important papers [16], [17], Wiles proved in most cases the so-called main conjecture for the cyclotomic $\mathbb{Z}_p$-extension of any totally real base field F, for all odd primes p, and for all abelian characters of $\mathrm{Gal}(\bar{\mathbb{Q}}/F)$ (we say most cases because his work only establishes the main conjecture up to μ-invariants for those abelian characters whose order is divisible by p). It would be technically too difficult for us in these introductory lectures to attempt to explain his proof in this generality. Instead, we have chosen the much more modest path of giving a sketch of his proof in the very special case that $F = \mathbb{Q}$, and for all abelian characters of $\mathbb{Q}$ of p-power conductor. In fact, our account of Wiles' proof in this case has been directly inspired by a series of lectures on this theme, which we attended, given by Chris Skinner in an instructional conference held at the Centre of Mathematical Sciences, Zhejiang University, Hangzhou, China, in August 2004. As Skinner's lectures were not written up, we thought it worthwhile to give here his account, insisting however that all inaccuracies in our version are of our own making. We remark that the proof as we have presented it here, makes use of the Ferrero-Washington theorem, asserting that $\mu = 0$ for the field obtained by adjoining all p-power roots of unity to $\mathbb{Q}$. However, we are grateful to C. Skinner for many helpful comments on our notes, including pointing put to us that a simple variant of the proof presented here avoids the use of the Ferrero-Washington theorem. We also stress that Wiles, in his work on the general case, had to overcome many addtional technical difficulties, for example the existence of the so-called trivial zeroes of p-adic L-function, and the fact that Leopoldt's conjecture is unknown for arbitrary totally real base fields. Moreover, at that time, it was not even known how to formulate the μ-invariant part of the main conjecture when the order of the character is divisible by p, whereas today we know how to do it in terms of K-theory. We also mention that other proofs of the main conjecture are known in this special case when $F = \mathbb{Q}$ (see [10], [14], [4]), but none of these generalise to base fields other than $\mathbb{Q}$. Finally, we thank the Department of Mathematics at POSTECH, South Korea, for providing us with excellent working conditions during the preparation of the written version of our lectures.

In what follows, F will denote a totally real number field, and p is an odd prime. We write Σ_p for the set of primes of F above p. As always, μ_{p^∞} is the group of all p-power roots of unity. The cyclotomic character is denoted by

$$\chi_F : \mathrm{Gal}(\bar{\mathbb{Q}}/F) \to \mathbb{Z}_p^\times,$$

so that $\sigma(\zeta) = \zeta^{\chi_F(\sigma)}$ for all σ in the Galois group and ζ in μ_{p^∞}. Let δ be the extension degree $[F(\mu_p) : F]$, and put $\Delta = \mathrm{Gal}(F(\mu_p)/F)$. The cyclotomic $\mathbb{Z}_p$-extension of F contained in $F(\mu_{p^\infty})$ is denoted by F^{cyc} and we put $\Gamma = \mathrm{Gal}(F^{\mathrm{cyc}}/F)$ so that $\Gamma \simeq \mathbb{Z}_p$. Recall that Leopoldt's conjecture for F is equivalent to the assertion that F^{cyc} is the unique $\mathbb{Z}_p$-extension of F. Finally for a complex variable s, $\zeta(F, s)$ will denote the complex zeta function of F, which is defined by

$$\zeta(F, s) = \prod_{v}(1 - (Nv)^{-s})^{-1}, \quad \mathrm{Re}(s) > 1.$$

In this first section, we state the "main conjecture" for our totally real number field F, but restricting our attention to powers of the Teichmüller character of F, rather than arbitrary abelian characters. Our starting point is the following theorem proved by Siegel [15].

Theorem 1.1 (Siegel). *For all even integers $n > 0$, we have $\zeta(F, 1 - n) \in \mathbb{Q}$.*

Siegel gave some nice formulae which allow one to compute these values by hand in some cases. For instance, if $F = \mathbb{Q}(\theta)$, where θ is a root of $x^3 - 9x + 1$, and F has discriminant 3×107. We have

$$\zeta(F, -1) = \pm 1, \quad \zeta(F, -3) = \pm\frac{3 \cdot 5 \cdot 37}{2}.$$

Definition 1.2. *Let G be any profinite group. The Iwasawa algebra $\Lambda(G)$ of G, is defined as*

$$\Lambda(G) = \varprojlim_{U}, \mathbb{Z}_p[G/U],$$

where U runs over all open normal subgroups of G.

The ring $\Lambda(G)$ has a second interpretation as the ring of $\mathbb{Z}_p$-valued measures on G. Thus, given any continuous function $f : G \to \mathbb{C}_p$, where $\mathbb{C}_p = \hat{\bar{\mathbb{Q}}}_p$, is the completion of the algebraic closure of $\mathbb{Q}_p$ for the p-adic topology, one can define the integral $\int_G f d\mu$ in $\mathbb{C}_p$.

2. p-adic L-functions

In this section, we state the theorem of Cassou-Noguès–Deligne–Ribet, (but only in the special case we need) in various equivalent forms. Define

$$\mathcal{F}_\infty = F(\mu_{p^\infty}), \quad F_\infty = F(\mu_{p^\infty})^+, \quad \mathcal{G} = \mathrm{Gal}(\mathcal{F}_\infty/F), \quad G = \mathrm{Gal}(F_\infty/F).$$

Definition 2.1. *An element μ of the fraction ring of $\Lambda(G)$ is a pseudo-measure if $(\sigma - 1)\mu \in \Lambda(G)$ for all σ in G.*

We also put

$$\zeta^{(p)}(F, s) = \zeta(F, s) \times \prod_{v|p}(1 - (Nv)^{-s}). \tag{1}$$

Two different proofs of the following theorem were given about the same time by Deligne–Ribet [5] and P. Cassou-Noguès [3].

Theorem 2.2 (Deligne–Ribet, Cassou-Nogués). *There exists a unique pseudo-measure $\mu_{F,p}$ on G such that, for all even integers $k > 0$, we have*

$$\int_G \chi_F^k \, d\mu_{F,p} = \zeta^{(p)}(F, 1-k).$$

There are several alternative ways of expressing this p-adic zeta function. We first explain the definition in terms of *branches*. We have

$$\mathrm{Gal}(\mathcal{F}_\infty/F) = \Delta \times \Gamma,$$

where Δ is isomorphic to $\mathrm{Gal}(F(\mu_p)/F)$ and Γ is isomorphic to $\mathrm{Gal}(F^{\mathrm{cyc}}/F)$. Put

$$\omega_F = \chi_{F|\Delta}, \quad \kappa_F = \chi_{F|\Gamma}.$$

It is traditional to call ω_F the Teichmüller character of F. The distinct characters of Δ are given by the ω_F^j $(j = 0, \ldots, \delta - 1)$. For s in $\mathbb{Z}_p$ and j even, the p-adic L-function $L_{F,p}(s, \omega_F^j)$ is defined as follows.

Definition 2.3. *For j even, $L_{F,p}(1 - s, \omega_F^j) = \int_G \omega_F^j \kappa_F^s \, d\mu_{F,p}$.*

It is immediate from the definition and the above theorem that, for all integers $k > 0$ with $k \equiv j \mod (p-1)$, we have

$$L_{F,p}(1-k, \omega_F^j) = \zeta^{(p)}(F, 1-k).$$

The method of proof of the theorem shows more generally that, for all integers $k > 0$, we have

$$L_{F,p}(1-k, \omega_F^j) = L^{(p)}(F, 1-k, \omega_F^{j-k}); \tag{2}$$

here $L(F, s, \omega_F^{j-k})$ denotes the *complex* L-function of ω_F^{j-k}, and $L^{(p)}(F, s, \omega_F^{j-k})$ means this complex L-function with the Euler factors of the primes dividing p removed.

There is yet another way of expressing these p-adic L-functions. We note that

$$\Lambda(G) = \mathbb{Z}_p[\Delta'][[\Gamma]], \quad \text{where} \quad \Delta' = \mathrm{Gal}(F(\mu_p)^+/F).$$

Let $e_{\omega_F^j}$ (j even) be the idempotent of ω_F^j in $\mathbb{Z}_p[\Delta']$. Then

$$\mathbb{Z}_p[\Delta'] = \bigoplus_{\substack{j \bmod \delta \\ j \text{ even}}} e_{\omega_F^j} \mathbb{Z}_p[\Delta'],$$

and evaluation at w_F^j defines a $\mathbb{Z}_p$-isomorphism from $e_{\omega_F^j}\mathbb{Z}_p[\Delta']$ to $\mathbb{Z}_p$. Thus there is a canonical decomposition

$$\Lambda(G) \simeq \bigoplus_{\substack{j \bmod \delta \\ j \text{ even}}} \Lambda(\Gamma).$$

Now fix a topological generator γ of Γ. There is a unique $\mathbb{Z}_p$-algebra isomorphism from $\Lambda(\Gamma)$ onto $\mathbb{Z}_p[[T]]$ which maps γ to $1+T$. Define

$$u = \chi_F(\gamma).$$

Hence the theorem of Cassou-Noguès–Deligne–Ribet can be stated as follows.

Theorem 2.4. *For each even integer j mod δ, there exists $W_{\omega_F^j}(T)$ in $\mathbb{Z}_p[[T]]$ such that*

$$L_{F,p}(s, \omega_F^j) = \begin{cases} W_{\omega_F^j}(u^s - 1) \text{ if } j \not\equiv 0 \bmod \delta \\ (W_{\omega_F^j}(u^s - 1))/(u^{1-s} - 1) \text{ if } j \equiv 0 \bmod \delta. \end{cases}$$

3. Statement of the "Main Conjecture"

We now give the statement of the "main conjecture" (Wiles' theorem) in various equivalent forms, but just for the even powers of ω_F. It turns out to be natural to consider two arithmetic $\mathcal{G}$-modules when one formulates this main conjecture. First, let M_∞ be the maximal abelian p-extension of F_∞, which is unramified outside the primes of F above p. Put

$$Y_\infty = \mathrm{Gal}(M_\infty/F_\infty). \tag{3}$$

Then Y_∞ is a compact $\mathbb{Z}_p$-module, which can be endowed with a continuous action of $G = \mathrm{Gal}(F_\infty/F)$ by defining $\sigma(y) = \tilde{\sigma} y \tilde{\sigma}^{-1}, \sigma \in G, y \in Y_\infty$), where $\tilde{\sigma}$ denotes any lifting of σ to $\mathrm{Gal}(M_\infty/F)$. This G-action extends by linearity and continuity to an action of the Iwasawa algebra $\Lambda(G)$. It is easy to see that Y_∞ is finitely generated over $\Lambda(G)$. A module N over $\Lambda(G)$ is said to be $\Lambda(G)$-torsion if every element of N is annihilated by a non-zero divisor in $\Lambda(G)$.

Theorem 3.1 (Iwasawa [8]). *The module Y_∞ is $\Lambda(G)$-torsion.*

For the second module, let $\mathcal{L}_\infty$ be the maximal unramified abelian p-extension of $\mathcal{F}_\infty = F(\mu_{p^\infty})$, and put

$$X_\infty = \mathrm{Gal}(\mathcal{L}_\infty/\mathcal{F}_\infty). \tag{4}$$

Analogously, X_∞ is also a module over the Iwasawa algebra $\Lambda(\mathcal{G})$. It is finitely generated over $\Lambda(\mathcal{G})$, and we have:

Theorem 3.2 (Iwasawa). *The module X_∞ is $\Lambda(\mathcal{G})$-torsion.*

Since p is odd, we can decompose X_∞ as

$$X_\infty = X_\infty^+ \oplus X_\infty^-,$$

where X_∞^ϵ ($\epsilon = \pm 1$) denotes the submodule of X_∞ on which the complex conjugation in $\mathcal{G}$ acts by ϵ. Very little is known about X_∞^+ (Greenberg has conjectured that X_∞^+ is finite for all F and all p, but no progress has been made towards the proof). However, there is a well-known connexion between the $\Lambda(\mathcal{G})$-modules Y_∞ and X_∞^-. If V is any $\Lambda(\mathcal{G})$-module, then we can change the action of $\mathcal{G}$ as follows. Firstly, we can define the new action of $\mathcal{G}$ by inverting the old action, i.e. defining $\sigma^{-1}(m)$ to be the new action of σ in $\mathcal{G}$ on m in V. We denote this new $\Lambda(\mathcal{G})$-module by $V^\bullet$. Secondly, we can twist by the inverse of the cyclotomic character, i.e. taking the new action of σ on m to be $\chi_F(\sigma)^{-1}\sigma(m)$. We denote this new $\Lambda(\mathcal{G})$-module by $V(-1)$. The proof of the following theorem uses both Kummer theory and the fact that all primes of F above p are ramified in F^{cyc}.

Theorem 3.3. *There is a $\Lambda(\mathcal{G})$-homomorphism from $(X_\infty^-)^\bullet$ to $Y_\infty(-1)$, with finite kernel and cokernel.*

Thanks to this theorem, we can state the "main conjecture" in two equivalent forms. As $G = \Delta' \times \Gamma$, where Δ' is cyclic of order dividing $p-1$, there is a classical structure theory for finitely generated torsion $\Lambda(G)$-modules (cf. [2], [4, Appendix]). It asserts that, if V is any finitely generated torsion $\Lambda(G)$-module, there exists an exact sequence of $\Lambda(G)$-modules

$$0 \to \bigoplus_{i=1}^{r} \Lambda(G)/f_i\Lambda(G) \to V \to D \to 0,$$

where $f_1, \ldots, f_r$ are non-zero divisors in $\Lambda(G)$, and D is finite. Note that the map on the left is injective because the module $\oplus_{i=1}^{r} \Lambda(G)/f_i\Lambda(G)$ has no non-zero finite $\Lambda(G)$-submodule. We then define the *characteristic ideal* $\mathrm{char}_G(V)$ of V by:

$$\mathrm{char}_G(V) = f_1 \ldots f_r\Lambda(G).$$

The kernel of the augmentation homomorphism $\Lambda(G) \to \mathbb{Z}_p$ is denoted by $I(G)$.

Theorem 3.4 ("Main Conjecture"-first version). *We have*

$$\mathrm{char}_G(Y_\infty) = \mu_{F,p}\, I(G),$$

where $\mu_{F,p}$ is the Deligne-Ribet–Cassou–Noguès pseudo-measure on G.

For the second version, we break the modules Y_∞ and X_∞ into eigenspaces for the action of $\Delta = \mathrm{Gal}(F(\mu_p)/F)$. If V is any $\Lambda(\mathcal{G})$-module, then we can decompose it as

$$V = \bigoplus_{j \bmod \delta} V^{(j)}, \quad \text{where} \quad V^{(j)} = e_{\omega_F^j} V.$$

We view each $V^{(j)}$ separately as a $\Lambda(\Gamma)$-module. Recall that we have fixed a topological generator γ of Γ, which gives rise to an isomorphism of rings

$$\Lambda(\Gamma) \simeq \mathbb{Z}_p[[T]].$$

Analogously to the above, we can define the characteristic ideal $\mathrm{char}_\Gamma(M)$ of any finitely generated torsion $\Lambda(\Gamma)$-module M. We recall that $W_{\omega_F^j}$ denotes the power series in $\mathbb{Z}_p[[T]]$ which gives the p-adic L-function $L_{F,p}(s, \omega_F^j)$ as explained in the last result of §2.

Theorem 3.5 ("Main Conjecture"-second version). *Let i be any odd integer mod δ. Then*

$$\mathrm{char}_\Gamma(X_\infty^{(-i)}) = W_{\omega_F^{i+1}}(T)\,\mathbb{Z}_p[[T]]. \tag{5}$$

To prove the equivalence of the two versions, we note that if $H_i(T)$ is a generator of the characteristic ideal of $X_\infty^{(-i)}$ (i odd), then $H_i(u(1+T)^{-1}-1)$ is a generator of the characteristic ideal of $Y_\infty^{(i+1)}$.

4. Overview of the proof of the main conjecture when $F = \mathbb{Q}$

In the remainder of these notes, we shall explain Wiles' proof of Theorem 3.5, but only in the special case when $F = \mathbb{Q}$. It is fair to say that the principles of the proof remain the same for an arbitrary totally real field F. Nevertheless, there are dramatic simplifications which occur when $F = \mathbb{Q}$ (for example Lemma 4.1 below does not have an elementary proof for fields other than $\mathbb{Q}$, nor do we know for such fields the analogue of the Ferrero-Washington theorem nor Leopoldt's conjecture). Also, for $F = \mathbb{Q}$, we need only work with elliptic modular forms rather than Hilbert modular forms.

We assume from now on that $F = \mathbb{Q}$, so that $\delta = p - 1$. We also simply write χ and ω rather than $\chi_\mathbb{Q}$ and $\omega_\mathbb{Q}$. Moreover, we choose γ to be the unique topological generator of Γ such that $\chi(\gamma) = 1 + p$. We have the classical formula

$$\zeta(\mathbb{Q}, 1-k) = -B_k/k \quad (k = 2, 4, 6\dots)$$

where the Bernoulli numbers are defined by the expansion

$$t/(e^t - 1) = \sum_{n=0}^{\infty} B_n t^n/n!.$$

We begin with a classical lemma, which enables us to discard two eigenspaces in the proof of Theorem 3.5 (and these eigenspaces cause technical difficulties in Wiles' proof).

Lemma 4.1. *If $i \equiv \pm 1$ mod $(p-1)$, we have that $X_\infty^{(-i)} = 0$ and $W_{\omega^i}(T)$ is a unit in $\mathbb{Z}_p[[T]]$. Thus the assertion (5) holds for these two eigenspaces.*

Proof. We first consider the case when $i \equiv -1$ mod $(p-1)$. The fact that $W_{\omega^0}(T)$ is a unit in $\mathbb{Z}_p[[T]]$ can be seen, for example, using the classical von Staudt Clausen theorem, which asserts that $\mathrm{ord}_p(B_{p-1}) = -1$. Then

$$W_{\omega^0}((1+p)^{2-p} - 1) = (1+p)^{p-1} - 1) \times L_{\mathbb{Q},p}(2-p, \omega^0) = \frac{B_{p-1}}{p-1}(1 - p^{p-2}).$$

Since $\mathrm{ord}_p((1+p)^{p-1}-1)=1$, it follows that $W_{\omega^0}((1+p)^{2-p}-1)$ is a p-adic unit, and hence $W_{\omega^0}(T)$ is a unit in $\mathbb{Z}_p[[T]]$. On the other hand, class field theory for $\mathbb{Q}$ proves that $\mathbb{Q}^{\mathrm{cyc}}$ is the maximal abelian p-extension of $\mathbb{Q}$ unramified outside p, whence it follows easily that $Y_\infty^0=0$, where Y_∞ is given by (3) above. Since Y_∞ is pseudo-isomorphic to $(X_\infty)^\bullet(1)$ by Theorem 3.3, we conclude that $X_\infty^{(1)}$ is finite. In fact, it is well-known that X_∞^- has no non-zero finite Γ-submodule, and so it follows that $X_\infty^{(1)}=0$. Next assume that $i\equiv 1 \bmod (p-1)$, and that $p\neq 3$. Then $\omega^2\neq\omega^0$, and we have

$$W_{\omega^2}((1+p)^{-1}-1)=L_{\mathbb{Q},p}(-1,\omega^2)=\frac{-B_2}{2}(1-p).$$

Since $B_2=1/6$, we see that $W_{\omega^2}((1+p)^{-1}-1)$ is a p-adic unit, and so $W_{\omega^2}(T)$ is a unit in $\mathbb{Z}_p[[T]]$. On the other hand, a classical argument in Iwasawa theory shows that

$$(X_\infty^{(-1)})_\Gamma=\mathcal{C}^{(-1)}, \tag{6}$$

where $\mathcal{C}$ denotes the p-primary subgroup of the class group of $\mathbb{Q}(\mu_p)$. Then, using Nakayama's lemma, it will follow from (6), that $X_\infty^{(-1)}=0$, provided we can show that $\mathcal{C}^{(-1)}=0$. To prove this last assertion, we note that the classical theorem of Stickelberger shows that $\mathcal{C}^{(-1)}$ is annihilated by $L(\mathbb{Q},0,\omega)$. However,

$$L(\mathbb{Q},0,\omega)=L_{\mathbb{Q},p}(0,\omega^2)=W_{\omega^2}(0),$$

and this last quantity is a p-adic unit because $W_{\omega^2}(T)$ is a unit in $\mathbb{Z}_p[[T]]$. This completes the proof of the lemma. □

From now on, i will be an odd integer mod $(p-1)$ satisfying

$$i\not\equiv -1 \bmod (p-1). \tag{7}$$

If d is in $\mathbb{Z}_p^\times$, we write $d=\omega(d)(1+p)^{e_d}$, with e_d in $\mathbb{Z}_p$.

The strategy for proving Theorem 3.5 is as follows. Let $\Lambda=\mathbb{Z}_p[[T]]$ be the ring of formal power series in T with coefficients in $\mathbb{Z}_p$. We recall that the Iwasawa algebra $\Lambda(\Gamma)$ is identified with Λ by mapping our canonical generator γ to $(1+T)$. In view of Lemma 4.1 and the analytic class number formula, it can be shown by a classical argument going back to Iwasawa that it suffices to prove that, for all odd integers i mod $p-1$, with $i\not\equiv\pm1$ mod $p-1$, we have

$$\mathrm{char}_\Gamma(X_\infty^{(-i)})\subset W_{\omega^{(i+1)}}(T)\Lambda. \tag{8}$$

The proof employs techniques from modular forms to prove this assertion, and we quickly give some indications now of the underlying ideas. Let $\mathcal{E}^{(i)}(T)$ be the element of the formal power series ring $\Lambda[[q]]$, which is defined by

$$\mathcal{E}^{(i)}(T)=A_0^{(i)}(T)+\sum_{n=1}^{\infty}A_n^{(i)}(T)q^n, \tag{9}$$

where

$$A_0^{(i)}(T) = W_{\omega^{i+1}}((1+T)^{-1}-1)/2, \quad \text{with} \quad L_{\mathbb{Q},p}(s,\omega^{i+1}) = W_{\omega^{i+1}}((1+p)^s-1), \tag{10}$$

and

$$A_n^{(i)}(T) = \sum_{\substack{d|n \\ (d,p)=1}} \omega^i(d)(1+T)^{e_d} \quad (n \geq 1);$$

here we are viewing ω as a Dirichlet character modulo p in the usual fashion, and as above, we have written $d = \omega(d)(1+p)^{e_d}$. This formal power series is the primaeval example of a Hida family (see [7]). For each integer $k \geq 2$, and any integer j mod $(p-1)$, let $\mathrm{M}_k(p,\omega^j,\mathbb{Z}_p)$ denote the space of classical modular forms of level p, character ω^j, and with coefficients in $\mathbb{Z}_p$. We recall that the Λ-module $\mathrm{M}(1,\omega^i,\Lambda)$ of Hida modular forms of level 1, character ω^i, with coefficients in Λ is defined to be the Λ-submodule of $\Lambda[[q]]$ consisting of f such that $\phi_k(f)$ belongs to $\mathrm{M}_k(p,\omega^{1+i-k},\mathbb{Z}_p)$, for all but a finite number of integers $k \geq 2$. Here ϕ_k is the $\mathbb{Z}_p$-algebra homomorphism from $\Lambda[[q]]$ to Λ defined by

$$\phi_k\left(\sum_{n=1}^{\infty} a_n(T)q^n\right) = \sum_{n=1}^{\infty} a_n((1+p)^{k-1}-1)q^n. \tag{11}$$

For each prime l, we have the Hecke operators T_l in $\mathrm{End}_\Lambda(\mathrm{M}(1,\omega^i,\mathbb{Z}_p))$, and also the operators S_l for all l (see [7]). In the next section, we shall give the proof of the following theorem, which is basic for all of our subsequent arguments.

Theorem 4.2. *Assume* (7). *Then $\mathcal{E}^{(i)}(T)$ belongs to* $\mathrm{M}(1,\omega^{\mathrm{i}},\Lambda)$. *Moreover, we have* $\mathrm{T}_p(\mathcal{E}^{(i)}(T)) = \mathcal{E}^{(i)}(T)$, *and for each prime* $l \neq p$,

$$\mathrm{T}_l(\mathcal{E}^{(i)}(T)) = (1+\omega^i(l)(1+T)^{e_l})\mathcal{E}^{(i)}(T),$$
$$\mathrm{S}_l(\mathcal{E}^{(i)}(T)) = \omega^i(l)l^{-1}(1+T)^{e_l}\mathcal{E}^{(i)}(T).$$

We recall that a classical modular form is said to be *ordinary* at p if its p-th Fourier coefficient is a p-adic unit. Let $\mathrm{M}^{\mathrm{ord}}(1,\omega^i,\Lambda)$ denote the space of ordinary Λ-adic modular forms of level 1 and character ω^i. By definition, it is the Λ-submodule consisting of all f in $\mathrm{M}(1,\omega^i,\Lambda)$ such that $\phi_k(f)$ is a classical ordinary modular form for all but a finite number of integers $k \geq 2$. An important theorem of Hida (see [7]) asserts that $\mathrm{M}^{\mathrm{ord}}(1,\omega^i,\Lambda)$ is a free Λ-module of finite rank. Since the Hecke operator T_p fixes $\mathcal{E}^{(i)}(T)$, it follows that $\mathcal{E}^{(i)}(T)$ belongs to $\mathrm{M}^{\mathrm{ord}}(1,\omega^i,\Lambda)$, and in view of the above theorem, this leads to the definition of the *Eisenstein ideal*, which plays a central role in Wiles' proof. Let $\mathbb{H}_i$ be the Λ-subalgebra of the endomorphism ring of $\mathrm{M}^{\mathrm{ord}}(1,\omega^i,\Lambda)$ which is generated by the identity, the Hecke operators T_l for all primes l, and by the S_l for all primes $l \neq p$.

Definition 4.3. *The Eisenstein ideal $\mathbb{I}_i$ is the ideal of $\mathbb{H}_i$ generated by* $\mathrm{T}_p - 1$ *and* $\mathrm{T}_l - 1 - \omega^i(l)(1+T)^{e_l}$, $\mathrm{S}_l - \omega^i(l)l^{-1}(1+T)^{e_l}$ *for all primes* $l \neq p$.

It is clear from Theorem 4.2 that $\mathcal{E}^{(i)}(T)$ is annihilated by the Eisenstein ideal $\mathbb{I}_i$.

Although the Eisenstein ideal is an ideal in the Hecke algebra $\mathbb{H}_i$, we need rather to work with the corresponding Hecke algebra for the ordinary Λ-adic cusp forms. Recall that $\mathrm{S}(1, \omega^i, \Lambda)$ denotes the Λ-submodule of $\mathrm{M}(1, \omega^i, \Lambda)$ consisting of all f such that, for all but a finite number of integers $k \geq 2$, $\phi_k(f)$ belongs to the space $\mathrm{S}_k(p, \omega^{i+1-k}, \mathbb{Z}_p)$ of classical cusp forms of level p, character ω^{i+1-k}, weight k, and with coefficients in $\mathbb{Z}_p$. Also, define

$$\mathrm{S}^{\mathrm{ord}}(1, \omega^i, \Lambda) = \mathrm{S}(1, \omega^i, \Lambda) \cap \mathrm{M}^{\mathrm{ord}}(1, \omega^{\mathrm{i}}, \Lambda).$$

Then the operators T_l (l any prime) and S_l (l any prime $\neq p$) leave $\mathrm{S}^{\mathrm{ord}}(1, \omega^i, \Lambda)$ stable. We can therefore define $\mathbb{T}_i$ to be the Λ-subalgebra of $\mathrm{End}_{\Lambda}(\mathrm{S}^{\mathrm{ord}}(1, \omega^i, \Lambda))$ which is generated by the identity, T_l (for all l) and the S_l (for all $l \neq p$). Moreover, restriction to the subspace $\mathrm{S}^{\mathrm{ord}}(1, \omega^i, \Lambda)$ clearly gives a canonical surjection from $\mathbb{H}_i$ onto $\mathbb{T}_i$, which allows us to view $\mathbb{T}_i$ as a $\mathbb{H}_i$-module.

The first key step in the proof of (8) is the following:

Theorem 4.4. *For all odd integers i with $i \not\equiv -1$ mod $(p-1)$, we have*

$$\mathrm{char}_{\Lambda}(\mathbb{T}_i/\mathbb{I}_i\mathbb{T}_i) \subset A_0^{(i)}(T)\Lambda. \tag{12}$$

To complete the proof of (8), one has to relate the Λ-module $\mathbb{T}_i/\mathbb{I}_i\mathbb{T}_i$ to the Λ-module $X_\infty^{(-i)}$. This is a beautiful and highly non-trivial argument, which seems to have been inspired by Ribet's work [13]. We explain it in some detail in the latter part of these notes. To carry it out, one is forced to enlarge the ring Λ, first by replacing Λ by the ring $\Lambda_{\mathcal{O}} = \mathcal{O}[[T]]$, where $\mathcal{O}$ is the ring of integers of some finite extension of $\mathbb{Q}_p$, and then by replacing $\Lambda_{\mathcal{O}}$ by the integral closure R of Λ in some finite extension of the fraction field of $\Lambda_{\mathcal{O}}$.

5. Eisenstein Families

The proof of Theorem 4.2 is entirely elementary, and we now give it. We continue to assume that i is an odd integer mod $(p-1)$, satisfying (7). Throughout, k will be an integer ≥ 2, and, if $f(T)$ is any element of Λ, we recall that $\phi_k(f) = f((1+p)^{k-1} - 1)$. The Weierstrass preparation theorem in Λ shows that $\phi_k(f) = 0$ for infinitely many integers $k \geq 2$ if and only if $f = 0$.

Put

$$\phi_k(\mathcal{E}^{(i)}(T)) = \sum_{n=0}^{\infty} a_n^{(k)} q^n.$$

From the definition of $\mathcal{E}^{(i)}(T)$, we have

$$\text{(i) } a_0^{(k)} = \phi_k(A_0^{(i)}(T)) = L_{\mathbb{Q},p}(1-k, \omega^{i+1})/2 = L^{(p)}(\mathbb{Q}, 1-k, \omega^{i+1-k})/2, \tag{13}$$

$$\text{(ii) } a_n^{(k)} = \phi_k(A_n^{(i)}(T)) = \sum_{\substack{d|n \\ (d,p)=1}} \omega^{i+1-k}(d) d^{k-1} \quad (n \geq 1). \tag{14}$$

We first show that $\mathcal{E}^{(i)}(T)$ belongs to $\mathrm{M}(1, \omega^i, \Lambda)$. Suppose first that $\omega^{i+1-k} \neq \mathbf{1}$, or equivalently that ω^{i+1-k} has conductor p. It follows that

$$a_0^{(k)} = 1/2\ L(\mathbb{Q}, 1-k, \omega^{i+1-k}), ca_n^{(k)} = \sum_{d|n} \omega^{i+1-k}(d)d^{k-1} \quad (n \geq 1).$$

Thus $\phi_k(\mathcal{E}^{(i)}(T))$ is the classical Eisenstein series of weight k, level p, and character ω^{i+1-k}. On the other hand, if $\omega^{i+1-k} = \mathbf{1}$, we see that

$$\phi_k(\mathcal{E}^{(i)}(T)) = E_k(z) - p^{k-1}E_k(pz)$$

where E_k is the classical Eisenstein series of weight k and level 1 given by

$$E_k = \zeta(\mathbb{Q}, 1-k)/2 + \sum_{n=1}^{\infty}\left(\sum_{d|n} d^{k-1}\right) q^n. \tag{15}$$

Hence in all cases, $\phi_k(\mathcal{E}^{(i)}(T))$ belongs to $\mathrm{M}_k(p, \omega^{i+1-k}, \mathbb{Z}_p)$, and this shows that $\mathcal{E}^{(i)}(T)$ is in $\mathrm{M}(1, \omega^i, \Lambda)$.

For simplicity, write T_l for both the classical and Λ-adic Hecke operator attached to a prime number l. To show that $\mathrm{T}_p\mathcal{E}^{(i)}(T) = \mathcal{E}^{(i)}(T)$, we must prove that

$$\mathrm{T}_p\, \phi_k(\mathcal{E}^{(i)}(T)) = \phi_k(\mathcal{E}^{(i)}(T))$$

for all integers $k \geq 2$. In other words, we must prove that $a_n^{(k)} = a_{np}^{(k)}$ for all integers $n \geq 1$. But this last equation is obvious from the explicit expression (ii) above for $a_n^{(k)}$ when $n \geq 1$.

Now assume that l is a prime different from p. By the defintion of the Hecke operator T_l, we have

$$\mathrm{T}_l(\mathcal{E}^{(i)}(T)) = \sum_{n=0}^{\infty} B_n^{(i)}(T)q^n,$$

where

$$B_n^{(i)}(T) = A_{nl}^{(i)}(T) + \delta_l\omega^i(l)\omega^i(l)(1+T)^{e_l}\, A_{n/l}^{(i)}(T);$$

here $\delta_l = 0$ or 1, according as $(l, n) = 1$ or $l|n$. Thus, to prove the assertion of Theorem 4.2 for the operator T_l, we must show that, for all $n \geq 0$, we have

$$B_n^{(i)}(T) = (1 + \omega^i(l)(1+T)^{e_l})\, A_n^{(i)}(T). \tag{16}$$

This last equation is obvious when $n = 0$, and so we may assume that $n > 0$. Suppose first that $(l, n) = 1$. In view of this last assertion, every positive divisor of nl is of the form j or jl, where j runs over all positive divisors of n. But, by definition, we then have

$$B_n^{(i)}(T) = A_{nl}^{(i)}(T) = \sum_{\substack{j|n \\ (j,p)=1}} \omega^i(j)(1+T)^{e_l} + \sum_{\substack{j|n \\ (j,p)=1}} \omega^i(jl)(1+T)^{e_{jl}}.$$

Since $e_{jl} = e_j + e_l$, the equation (16) follows immediately in this case. Next suppose that $l|n$. Since $\delta_l = 1$ in this case, we see immediately that

$$B_n^{(i)}(T) = \sum_{\substack{d|nl \\ (d,p)=1}} \omega^i(d)(1+T)^{e_d} + \sum_{\substack{jl|n \\ (j,p)=1}} \omega^i(jl)(1+T)^{e_{jl}}.$$

Now we clearly have

$$\sum_{\substack{d|nl \\ (d,p)=1}} \omega^i(d)(1+T)^{e_d} = \sum_{\substack{d_1|n \\ (d_1,p)=1}} \omega^i(d_1)(1+T)^{e_{d_1}} + \sum_{\substack{d_2|n \\ (d_2,p)=1 \\ d_2\nmid n}} \omega^i(d_2)(1+T)^{e_{d_2}}.$$

Since

$$\sum_{\substack{b|n \\ (b,p)=1}} \omega^i(bl)(1+T)^{e_{bl}} = \sum_{\substack{jl|n \\ (j,p)=1}} \omega^i(jl)(1+T)^{e_{jl}} + \sum_{\substack{d_2|nl \\ (d_2,p)=1 \\ d_2\nmid n}} \omega^i(d_2)(1+T)^{e_{d_2}},$$

we see that we have again established equation (16). The proof of the final assertion of Theorem 4.2 is straightforward and we omit the details.

6. Ribet's theorem

Our aim in this section is to give a proof of Ribet's theorem, which will serve as an introduction to Wiles' proof [17] of the Main Conjecture (Theorem 3.4).

We continue to assume that our base field $F = \mathbb{Q}$, and we put $K = \mathbb{Q}(\mu_p)$. As earlier, $\Delta = \mathrm{Gal}(K/\mathbb{Q})$, and ω is the character giving the action of Δ on μ_p. Let $\mathcal{C}$ denote the p-primary subgroup of the ideal class group of K. Then, as before, we have the decomposition

$$\mathcal{C} = \bigoplus_{i \bmod (p-1)} \mathcal{C}^{(i)},$$

where $\mathcal{C}^{(i)} = e_{\omega^i}\mathcal{C}$.

Theorem 6.1 (Ribet [13]). *Assume k is an even number with $2 \le k \le p-3$. If an odd prime p divides the numerator of $\zeta(\mathbb{Q}, 1-k)$, then $\mathcal{C}^{(1-k)} \neq 0$.*

We are grateful to Skinner for communicating to us the following elegant variant of Ribet's original proof. We may assume that $p > 7$, since

$$\zeta(\mathbb{Q}, -1) = -1/12, \quad \zeta(\mathbb{Q}, -3) = 1/120.$$

Also,

$$\zeta(\mathbb{Q}, -5) = -1/252.$$

Hence the classical Eisenstein series E_4 and E_6 (see (15)) have p-adic integral Fourier expansions, with constant terms which are p-adic units. As k is even, we can find integers a, $b \ge 0$ so that $k = 4a + 6b$. Put $g = E_4^a E_6^b$, and define

$$h = E_k - (a_0(g)^{-1}\zeta(\mathbb{Q}, 1-k)/2)g,$$

where $a_0(g)$ denotes the constant term of the q-expansion of g. In general, if f is any classical modular form, we write $\sum_{n=0}^{\infty} a_n(f)q^n$, for its Fourier expansion.

Since p is assumed to divide the numerator of $\zeta(\mathbb{Q}, 1-k)$, we conclude that

$$a_n(h) \equiv a_n(E_k) \bmod p \ (n \geq 1).$$

In particular, we obtain

$$\begin{aligned} a_p(h) &\equiv 1 + p^{k-1} \equiv 1 \bmod p \\ a_l(h) &\equiv 1 + l^{k-1} \equiv 1 + \omega^{k-1}(l) \bmod p \quad (l \neq p). \end{aligned} \tag{17}$$

Let $\mathrm{S}_k^{\mathrm{ord}}(p, \mathbb{Z}_p)$ denote the space of ordinary cusp forms of weight k, level p and coefficients in $\mathbb{Z}_p$. The Hecke operators T_l (l any prime) and S_l (l any prime $\neq p$) leave $\mathrm{S}_k^{\mathrm{ord}}(p, \mathbb{Z}_p)$ stable, and we define $\mathbb{T}_{k,p}$ to be the $\mathbb{Z}_p$-subalgebra of $\mathrm{End}_{\mathbb{Z}_p}(\mathrm{S}_k^{\mathrm{ord}}(p, \mathbb{Z}_p))$ generated by these operators and the identity endomorphism. Since $\mathrm{S}_k^{\mathrm{ord}}(p, \mathbb{Z}_p)$ is a free $\mathbb{Z}_p$-module of finite rank, $\mathbb{T}_k$ is also a free $\mathbb{Z}_p$-module of finite rank. If t is in $\mathbb{T}_{k,p}$, we shall follow classical notation and write $f|t$ for the image under t of an element f of $\mathrm{S}_k^{\mathrm{ord}}(p, \mathbb{Z}_p)$. We now define a $\mathbb{Z}_p$-algebra homomorphism

$$\pi_k : \mathbb{T}_{k,p} \to \mathbb{F}_p$$

by

$$\pi_k(t) = a_1(h|t) \bmod p. \tag{18}$$

Let $\mathcal{M} = \mathrm{Ker}\,(\pi_k)$. Since (17) shows that, for all primes $l \neq p$, we have

$$\pi_k(\mathrm{T}_l) \equiv a_l(h) \equiv 1 + \omega^{k-1}(l) \bmod p, \tag{19}$$

we see immediately that, for all $l \neq p$,

$$\mathrm{T}_l - 1 - \omega^{k-1}(l) \in \mathcal{M}. \tag{20}$$

Take $\mathfrak{p}$ to be any minimal prime of $\mathbb{T}_{k,p}$ which is contained in $\mathcal{M}$. Since $\mathbb{T}_{k,p}/\mathfrak{p}$ is a commutative ring that is free of finite rank as a $\mathbb{Z}_p$-module, it has Krull dimension 1. We can therefore find an injective ring homomorphisn

$$\psi : \mathbb{T}_{k,p}/\mathfrak{p} \hookrightarrow \bar{\mathbb{Q}}_p.$$

Write $\mathcal{O}$ for the ring of integers of the field of fractions of the image of ψ. Clearly $\mathcal{O}$ has Krull dimension one as $\mathfrak{p}$ is minimal. By the duality between Hecke algebras and modular forms (see for example, [1, 6.5]), it is a classical result (see [6, pp. 325–328]) that the element $\psi \in \mathrm{Hom}_{\mathcal{O}}(\mathbb{T}_{k,p}/\mathfrak{p}, \mathcal{O})$ corresponds to a primitive eigenform $f \in \mathrm{S}_k^{\mathrm{ord}}(p, \mathcal{O})$ such that

$$\mathrm{T}_l(f) = a_l(f) \cdot f$$

where $a_l(f) = \psi(T_l)$ for all primes l. On the other hand, if π is a uniformiser of $\mathcal{O}$, then

$$a_l(f) \equiv 1 + \omega^{k-1}(l) \bmod \pi, \ l \neq p, \quad \text{and} \quad a_p(f) \equiv 1 \bmod \pi. \tag{21}$$

Since f is a primitive eigenform, there is an associated irreducible Galois representation (see [1, §6])

$$\rho = \rho_f : G_{\mathbb{Q}} \to \mathrm{GL}_2(K)$$

where K is the fraction field of $\mathcal{O}$. Choosing a lattice stable under $G_{\mathbb{Q}}$, we may assume that

$$\rho : G_{\mathbb{Q}} \to \mathrm{GL}_2(\mathcal{O}).$$

Let D_p denote the decomposition group at p. By the hypothesis that f is ordinary, we have (see [11])

$$\rho_{|D_p} \simeq \begin{pmatrix} \psi_1 & * \\ 0 & \psi_2 \end{pmatrix}.$$

where ψ_2 is an unramified character with $\psi_2(\mathrm{Frob}_p) = a_p(f)$, and Frob_p is the Frobenius at p. By conjugating with the element $\left(\begin{smallmatrix} 0 & 1 \\ 1 & 0 \end{smallmatrix}\right)$, we may write

$$\rho_{|D_p} = \begin{pmatrix} \psi_2 & 0 \\ * & \psi_1 \end{pmatrix}. \tag{22}$$

Further, as ρ is unramified outside p, we have

$$\det(\rho(\mathrm{Frob}_l)) = l^{k-1}, \quad l \neq p,$$

and hence

$$\det(\rho) = \chi^{k-1},$$

where, as before, $\chi = \chi_{\mathbb{Q}}$ is the cyclotomic character. Thus, as $\psi_{2|I_p} = 1$, we have

$$\psi_{1|I_p} = \det \rho_{|I_p} = \chi^{k-1}_{|I_p}.$$

But $\omega = \chi \bmod \pi$, and hence

$$\psi_{1|I_p} \equiv \omega^{k-1} \bmod \pi, \quad \text{and } \omega^{k-1} \not\equiv 1 \bmod \pi \text{ as } k \text{ is even.}$$

We therefore conclude that there exists an element $\sigma_0 \in I_p$ such that $\psi_2(\sigma_0) = 1$ and $\psi_1(\sigma_0) = \alpha$, with $\alpha \in \mathcal{O}^\times$ and $\alpha \not\equiv 1 \bmod \pi$. As $\rho(\sigma_0)$ has distinct eigenvalues 1 and α, we can choose a new $\mathcal{O}$-basis for the representation ρ such that (22) remains valid and we have

$$\rho(\sigma_0) = \begin{pmatrix} 1 & 0 \\ 0 & \alpha \end{pmatrix}, \tag{23}$$

For σ in $\mathrm{Gal}(\bar{\mathbb{Q}}/\mathbb{Q})$, write

$$\rho(\sigma) = \begin{pmatrix} a_\sigma & b_\sigma \\ c_\sigma & d_\sigma \end{pmatrix}.$$

By (22), we have $b_\sigma = 0$ for σ in I_p, and as ρ is unramified outside p, $b_\sigma = 0$ for σ in I_l $(l \neq p)$. Further, as ρ is irreducible, there exists an element τ in $\mathrm{Gal}(\bar{\mathbb{Q}}/\mathbb{Q})$

such that $b_\tau \neq 0$. Let τ_0 be an element in $\mathrm{Gal}(\bar{\mathbb{Q}}/\mathbb{Q})$ with the property that $\mathrm{ord}_\pi(b_{\tau_0})$ is minimal among all such choices of τ, and put $n = \mathrm{ord}_\pi(b_{\tau_0})$. Replacing ρ by

$$\begin{pmatrix} 1 & 0 \\ 0 & \pi^n \end{pmatrix} \rho \begin{pmatrix} 1 & 0 \\ 0 & \pi^{-n} \end{pmatrix},$$

we may assume that $n = 0$, in other words, that b_{τ_0} belongs to $\mathcal{O}^\times$. Put $\mathbb{F} = \mathcal{O}/\pi\mathcal{O}$ and let

$$\bar{\rho} : \mathrm{Gal}(\bar{\mathbb{Q}}/\mathbb{Q}) \to \mathrm{GL}_2(\mathbb{F})$$

be the residual representation of ρ mod π. By (21), we have

$$\mathrm{Trace}(\bar{\rho}) = 1 + \omega^{k-1}; \quad \det(\bar{\rho}) = \omega^{k-1}. \tag{24}$$

Let $\bar{\rho}^{\mathrm{ss}}$ be the semi-simplification of the residual representation $\bar{\rho}$. Since $\bar{\rho}^{\mathrm{ss}}$ and $\mathbf{1} \oplus \omega^{k-1}$ are two semi-simple representations over $\mathbb{F}$ with the same characteristic polynomials, thanks to (24), we conclude from the Brauer-Nesbitt theorem, that

$$\bar{\rho}^{\mathrm{ss}} = \mathbf{1} \oplus \omega^{k-1}.$$

From (22) and (23),and the fact that ω^{k-1} is ramified at p, we conclude that $\bar{\rho}$ is of the form

$$\begin{pmatrix} 1 & 0 \\ 0 & \omega^{k-1} \end{pmatrix}, \quad \begin{pmatrix} 1 & 0 \\ * & \omega^{k-1} \end{pmatrix}, \quad \text{or} \quad \begin{pmatrix} 1 & * \\ 0 & \omega^{k-1} \end{pmatrix}, \tag{25}$$

with the second two possibilities non-split. From the non-vanishing of $\bar{b}_{\tau_0}$, we see that it must be the third possibility. In other words, we have

$$\bar{\rho}(\sigma) = \begin{pmatrix} 1 & \bar{b}_\sigma \\ 0 & \omega^i(\sigma) \end{pmatrix}$$

for all σ in $\mathrm{Gal}(\bar{\mathbb{Q}}/\mathbb{Q})$. Let $\mathbb{F}(\omega^{1-k})$ denote the $\mathbb{F}$-vector space of dimension one on which $\mathrm{Gal}(\bar{\mathbb{Q}}/\mathbb{Q})$ acts via the character ω^{1-k}. Define a map h in $H^1(G_\mathbb{Q}, \mathbb{F}(\omega^{1-k}))$ by

$$h : \mathrm{Gal}(\bar{\mathbb{Q}}/\mathbb{Q}) \to \mathbb{F}(\omega^{1-k})$$
$$\sigma \mapsto \omega^{1-k}(\sigma)\bar{b}_\sigma.$$

As

$$b_{\sigma\tau} = b_\tau + b_\sigma\omega^{k-1}(\tau),$$

it follows that h is a 1-cocycle on $\mathrm{Gal}(\bar{\mathbb{Q}}/\mathbb{Q})$ with values in $\mathbb{F}(\omega^{1-k})$. Moreover, h is an unramified cocycle because $b_\sigma = 0$ for all σ in I_p, and also in I_l, for all $l \neq p$ any prime, because the representation ρ is unramified outside p. Thus we obtain an unramified 1-cocycle in $\bar{h}$ in $H^1(\mathrm{Gal}(\bar{\mathbb{Q}}/\mathbb{Q}, \mathbb{F}(\omega^{1-k}))$. On the other hand, as $\Delta = \mathrm{Gal}(\mathbb{Q}(\mu_p)/\mathbb{Q})$ has order prime to p, we see that

$$H^1(\mathrm{Gal}(\bar{\mathbb{Q}}/\mathbb{Q}, \mathbb{F}(\omega^{1-k})) = \mathrm{Hom}_\Delta(\mathrm{Gal}(\bar{\mathbb{Q}}/\mathbb{Q}(\mu_p)), \mathbb{F}(\omega^{1-k})).$$

Moreover, as $\bar{b}_{\tau_0}$ is not zero, it is plain that $\bar{h}$ is non-zero. Hence, by class field theory, we must have $\mathrm{Hom}_\Delta(\mathcal{C}^{(1-k)}, \mathbb{F}(\omega^{1-k}))$ is not equal to zero. Thus $\mathcal{C}^{(1-k)} \neq 0$, thereby proving Ribet's theorem. □

7. Λ-adic setting for the main conjecture

The aim of this section is to prove Theorem 4.4, and thus we fix for the rest of this section an odd integer i with $i \not\equiv -1 \bmod (p-1)$.

Let $\mathcal{O}$ be the ring of integers of a finite extension of $\mathbb{Q}_p$, and $\Lambda_{\mathcal{O}}$ the ring of formal power series in T with coefficients in $\mathcal{O}$. We can then consider the spaces $\mathrm{M}^{\mathrm{ord}}(1, \omega^i, \Lambda_{\mathcal{O}}) \supseteq \mathrm{S}^{\mathrm{ord}}(1, \omega^i, \Lambda_{\mathcal{O}})$, and define the objects $\mathbb{H}_i$, $\mathbb{T}_i$, $\mathbb{I}_i$ made with Λ replaced by $\Lambda_{\mathcal{O}}$. It is then very easy to see that it suffices to prove the assertion in Theorem 4.4 with Λ replaced by $\Lambda_{\mathcal{O}}$. Thus, from now on, we shall work with the objects formed with $\Lambda_{\mathcal{O}}$, with $\mathcal{O}$ sufficiently large.

We must invoke the Ferrero-Washington theorem, which asserts that p does not divide the power series $A_0^{(i)}(T)$. Thus, taking $\mathcal{O}$ sufficiently large, we can write

$$A_0^{(i)}(T) = \prod_{r=1}^{n_i} (T - \alpha_j^{(i)}) B_0^{(i)}(T),$$

where all of the roots $\alpha_j^{(i)}$ belong to $\mathcal{O}$, and $B_0^{(i)}(T)$ is a unit in $\Lambda_{\mathcal{O}}$. Choose an integer $m > 0$ with $m \equiv 0 \bmod (p-1)$ such that

$$\{\alpha_j^{(i)} : j = 1, \ldots, n_i\} \cap \{(\alpha_j^{(i)} + 1)(1+p)^m - 1 : j = 1, \ldots, n_i\} = \emptyset. \tag{26}$$

We now define

$$H = \mathcal{E}^{(i)}((1+T)(1+p)^{-m} - 1) \cdot E_m.$$

Put

$$H = \sum_{n=0}^{\infty} a_0(H) q^n.$$

Lemma 7.1. *H belongs to $\mathrm{M}(1, \omega^i, \Lambda_{\mathcal{O}})$. Moreover, the formal power series $a_0(H)$ has no zero in common with $A_0^{(i)}(T)$.*

Proof. We deduce from (9) with $i = 1$ that

$$a_0(H) = A_0^{(i)}((1+T)(1+p)^{-m} - 1) \cdot \zeta(\mathbb{Q}, 1-m)/2,$$

and conclude from (26) that $a_0(H)$ has no zero in common with $A_0^{(i)}(T)$.

Since $\omega^m = 1$, we see immediately that $\phi_k(H)$ belongs to $\mathrm{M}_k(p, \omega^{i+1-k}, \mathbb{Z}_p)$ for all sufficiently large integers k. Thus H belongs to $\mathrm{M}(1, \omega^i, \Lambda_{\mathcal{O}})$ and the proof of the lemma is complete. □

Recall the Hida operator $\mathfrak{e}$ on $\mathrm{M}(1, \omega^i, \Lambda_{\mathcal{O}})$ defined by the formula

$$\mathfrak{e} = \lim_{n \to \infty} \mathrm{T}_p^{n!}. \tag{27}$$

where T_p is now the Hecke operator of p on $\mathrm{M}(1, \omega^i, \Lambda_{\mathcal{O}})$. We have [7]

$$\mathrm{M}^{\mathrm{ord}}(1, \omega^i, \Lambda_{\mathcal{O}}) = \mathfrak{e}\mathrm{M}(1, \omega^i, \Lambda_{\mathcal{O}}), \quad \mathrm{S}^{\mathrm{ord}}(1, \omega^i, \Lambda_{\mathcal{O}}) = \mathfrak{e}\mathrm{S}(1, \omega^i, \Lambda_{\mathcal{O}}).$$

We then make the crucial definition

$$H' = \mathfrak{e}\,(a_0(H)\mathcal{E}^{(i)} - A_0^{(i)} H), \tag{28}$$

so that H' belongs to $\mathrm{M}^{\mathrm{ord}}(1, \omega^i, \Lambda_{\mathcal{O}})$.

Lemma 7.2. *In fact, H' belongs to $\mathrm{S}^{\mathrm{ord}}(1, \omega^i, \Lambda_{\mathcal{O}})$.*

Proof. By construction $a_0(H') = 0$. The assertion then follows from a general theorem of Wiles (see [16] and [1, §7]). □

Now let $\mathfrak{p} = (T - \alpha_j^{(i)})\Lambda_{\mathcal{O}}$ for some j, and put $m = \mathrm{ord}_{\mathfrak{p}}(A_0^{(i)}(T))$. It follows from Lemma 7.1 that $\mathrm{ord}_{\mathfrak{p}}(a_0(H)) = 0$, or equivalently that $a_0(H)^{-1}$ belongs to $\Lambda_{\mathcal{O},\mathfrak{p}}$, the localisation of $\Lambda_{\mathcal{O}}$ at $\mathfrak{p}$. The Hecke algebra

$$\mathbb{T}_{i,\mathfrak{p}} = \mathbb{T}_i \otimes_{\Lambda_{\mathcal{O}}} \Lambda_{\mathcal{O},\mathfrak{p}}$$

acts on

$$\mathrm{S}^{\mathrm{ord}}(1, \omega^i, \Lambda_{\mathcal{O}}) \otimes_{\Lambda_{\mathcal{O}}} \Lambda_{\mathcal{O},\mathfrak{p}},$$

and we define the $\Lambda_{\mathcal{O},\mathfrak{p}}$-homomorphism

$$\Pi_i : \mathbb{T}_{i,\mathfrak{p}} \to \Lambda_{\mathcal{O},\mathfrak{p}}/\mathfrak{p}^m$$

by

$$\Pi_i(t) = a_0(H)^{-1} \cdot a_1(H'|t) \bmod \mathfrak{p}^m.$$

Lemma 7.3. *We have $\Pi_i(t) = a_1(\mathcal{E}^{(i)}|t) \bmod \mathfrak{p}^m$. In particular $\mathbb{I}_i$ is contained in the kernel of Π_i. Also, Π_i is surjective.*

Proof. We have $\mathfrak{e}\,\mathcal{E}^{(i)} = \mathcal{E}^{(i)}$, and so

$$H' = a_0(H)\mathcal{E}^{(i)} - A_0^{(i)}\,\mathfrak{e}\,H.$$

Thus

$$H'|t = a_0(H)(\mathcal{E}^{(i)}|t) - A_0^{(i)}(\mathfrak{e}\,H|t).$$

Since $A_0^{(i)} \cdot \Lambda_{\mathcal{O}} = \mathfrak{p}^m$, the first assertion of the lemma follows. The second assertion is then clear because $\mathbb{I}_i$ annihilates $\mathcal{E}^{(i)}$. Finally, the surjectivity of Π_i is plain because $\mathbb{T}_i$ contains the identity endomorphism. □

In view of the above lemma, we see that there is a surjective homomorphism

$$\mathbb{T}_{i,\mathfrak{p}}/\mathbb{I}_i\mathbb{T}_{i,\mathfrak{p}} \to \Lambda_{\mathcal{O},\mathfrak{p}}/\mathfrak{p}^m.$$

As this is true for every root of $A_0^{(i)}(T)$, it follows that

$$\mathrm{char}_{\Lambda_{\mathcal{O}}}(\mathbb{T}_i/\mathbb{I}_i\mathbb{T}_i) \subseteq A_0^{(i)}(T)\Lambda_{\mathcal{O}}.$$

This completes the proof of Theorem 4.4. □

8. First part of the proof of the main conjecture

We continue to assume that i is an odd integer with $i \not\equiv -1 \bmod (p-1)$. Recall that our goal is to prove that

$$\mathrm{char}_{\Lambda}(X_{\infty}^{(-i)}) \subset A_0^{(i)}((1+T)^{-1}-1)\Lambda, \tag{29}$$

since, as remarked earlier (see Lemma 4.1, (8) and (10)), this implies the main conjecture (Theorem 3.5) by a classical argument due to Iwasawa [9].

In order to carry out our subsequent arguments, we have to enlarge the ring $\Lambda_{\mathcal{O}}$. The reason for this is the following. Let $\mathfrak{q}_1, \ldots, \mathfrak{q}_s$ be the minimal prime ideals of the Hecke algebra $\mathbb{T}_i$ defined using $\Lambda_{\mathcal{O}}$. By a theorem of Hida [7, Theorem 7, Chap. 7, p. 221], all of the quotient rings $\mathbb{T}_1/\mathfrak{q}_j$ $(j = 1, \ldots, s)$ are contained in the integral closure R of $\Lambda_{\mathcal{O}}$ in a suitable finite extension of the quotient field of $\Lambda_{\mathcal{O}}$. We shall henceforth consider the spaces

$$\mathrm{M}^{\mathrm{ord}}(1, \omega^i, R) \supseteq \mathrm{S}^{\mathrm{ord}}(1, \omega^i, R),$$

and the endomorphism algebras $\mathbb{H}_i$, $\mathbb{T}_i$, and the Eisenstein ideal $\mathbb{I}_i$, made by replacing $\Lambda_{\mathcal{O}}$ by the ring R. It follows that for this new Hecke algebra $\mathbb{T}_i$ with respect to the ring R, we have ring isomorphisms

$$\eta_j : \mathbb{T}_i/\mathfrak{q}_j \simeq R \quad (1 \leq j \leq s) \tag{30}$$

for every minimal prime ideal $\mathfrak{q}_j$ of $\mathbb{T}_i$. By the usual duality (see [7, Theorem 5, Chap. 7]), there then exists unique normalized ordinary eigenforms

$$f_j = \sum_{k=1}^{\infty} c_k(f_j) q^k \ (1 \leq j \leq s),$$

whose Fourier coefficients $c_k(f_j)$ $(k \geq 1)$ are in R, such that

$$\eta_j(\mathrm{T}_l) = c_l(f_j) \tag{31}$$

for all prime numbers l. We remark that each homomorphism ϕ_k on $\Lambda_{\mathcal{O}}$ has a finite number of extensions to the ring R, and each of the f_j's can be specialised under all these extensions to classical modular forms [7].

It is clear from the definition of the Eisenstein ideal that there is a natural surjection as R-algebras from R onto $\mathbb{T}_i/\mathbb{I}_i\mathbb{T}_i$, giving rise to an isomorphism

$$\lambda_i : R/J_i \simeq \mathbb{T}_i/\mathbb{I}_i\mathbb{T}_i \tag{32}$$

for some ideal J_i of R. We now fix for the rest of this section a prime ideal $\mathcal{P}$ of height one in R. Exactly the same arguments as those used to prove Theorem 4.4 show that again we have

$$\mathrm{ord}_{\mathcal{P}}(\mathrm{char}_R(\mathbb{T}_i/\mathbb{I}_i\mathbb{T}_i)) \geq \mathrm{ord}_{\mathcal{P}}(A_0^{(i)}(T)); \tag{33}$$

here for any finitely generated torsion R-module M, $\mathrm{char}_R(M)$ denotes its characteristic ideal.

The next important step in the proof of the main conjecture is to relate

$$\mathrm{ord}_{\mathcal{P}}(\mathrm{char}_R(R/J_i)) = \mathrm{ord}_{\mathcal{P}}(\mathrm{char}_R(\mathbb{T}_i/\mathbb{I}_i\mathbb{T}_i)) \tag{34}$$

to $\mathrm{char}_R(X_\infty^{(-i)} \otimes_\Lambda R)$ by an elaboration to the R-adic setting of Ribet's arguments, as set out in §6. First, we note that we may assume from now on that

$$\mathrm{ord}_{\mathcal{P}}(A_0^{(i)}) > 0,$$

since otherwise there is nothing to prove in (29). In view of (32), it follows that $\mathcal{P} \supset J_i$.

Put

$$\mathbb{T}_{i,\mathcal{P}} = \mathbb{T}_i \otimes_R R_{\mathcal{P}},$$

where $R_{\mathcal{P}}$ is the localization of R at $\mathcal{P}$. Write $\mathfrak{q}_1, \ldots, \mathfrak{q}_n$ for the minimal prime ideals of the localized Hecke algebra $\mathbb{T}_{i,\mathcal{P}}$. By a result of Hida [7, Theorem 4, Chap. 7], the algebra $\mathbb{T}_i$, and therefore also its localization $\mathbb{T}_{i,\mathcal{P}}$ has no nilpotent elements. Thus the natural R-algebra map

$$\mathbb{T}_{i,\mathcal{P}} \to \prod_{j=1}^{n} \mathbb{T}_{i,\mathcal{P}}/\mathfrak{q}_j \simeq \prod_{j=1}^{n} R_{\mathcal{P}} \tag{35}$$

is injective. Note also that $R_{\mathcal{P}}$ is an integrally closed, local, Noetherian domain of Krull dimension one, and hence is a discrete valuation ring.

Let us suppose that the indices are chosen so that the normalized ordinary eigenforms $f_1, \ldots, f_n$ correspond via (31) to those minimal prime ideals of $\mathbb{T}_i$ which, via localization give $\mathfrak{q}_1, \ldots, \mathfrak{q}_n$ in $\mathbb{T}_{i,\mathcal{P}}$. Thanks to Hida (see [7, §7.5, Chap. 7]), there exist irreducible Galois representations

$$\rho_j : \mathrm{Gal}(\bar{\mathbb{Q}}/\mathbb{Q}) \to \mathrm{Aut}(V_j), \ (j \le 1 \le n) \tag{36}$$

where the V_j are vector spaces of dimension two over the fraction field of R, such that ρ_j is unramified outside of p and

$$\mathrm{Trace}\, \rho_j(\mathrm{Frob}_l) = c_l(f_j) \qquad (l \ne p). \tag{37}$$

Moreover, we have

$$\det \rho_j = \theta_i,$$

where

$$\theta_i : \mathrm{Gal}(\bar{\mathbb{Q}}/\mathbb{Q}) \to R^\times$$

is the unique homomorphism which factors through the quotient $\Delta \times \Gamma$, and is defined on this quotient by

$$\theta_{i|_\Delta} = \omega^i, \quad \theta_i(\gamma) = 1 + T, \tag{38}$$

where, as before, γ is our fixed topological generator of Γ.

Let D_p denote the decomposition group of a fixed prime of $\bar{\mathbb{Q}}$ above p and I_p its inertial subgroup. Since $f_1, \ldots, f_n$ are ordinary eigenforms, a theorem of Wiles [16] asserts that we can find an $R_{\mathcal{P}}$-basis of M_j such that the restriction of ρ_j to D_p relative to this basis is of the form

$$\rho_{j|D_p} = \begin{pmatrix} \Psi_1^{(j)} & * \\ 0 & \Psi_2^{(j)} \end{pmatrix}$$

where $\Psi_i^{(j)}$ and $\Psi_2^{(j)}$ are homomorphisms from D_p to $R_{\mathcal{P}}^{\times}$, with $\Psi_2^{(j)}{}_{|I_p} = \mathbf{1}$. Conjugating by the matrix $\left(\begin{smallmatrix} 0 & 1 \\ 1 & 0 \end{smallmatrix}\right)$, we may assume that

$$\rho_{j|D_p} = \begin{pmatrix} \Psi_2^{(j)} & 0 \\ * & \Psi_1^{(j)} \end{pmatrix}. \tag{39}$$

We make one further modification of our $R_{\mathcal{P}}$-basis of M_j. Since p is totally ramified in $\mathbb{Q}(\mu_{p^\infty})$, we can fix a lifting $\tilde{\gamma}$ of γ to I_p whose image in Δ is trivial. As $\Psi_2^{(j)}(\tilde{\gamma}) = 1$, and

$$\Psi_2^{(j)}\Psi_1^{(j)} = \det\, \rho_{j|D_p} = \theta_{i|D_p},$$

we obtain $\Psi_1^{(j)}(\tilde{\gamma}) = 1 + T$. Since $\rho_j(\tilde{\gamma})$ has the distinct eigenvalues 1 and $1+T$, it is easy to see that we can choose a new basis of M_j such that (39) still remains true and, in addition,

$$\rho_j(\tilde{\gamma}) = \begin{pmatrix} 1 & 0 \\ 0 & 1+T \end{pmatrix}. \tag{40}$$

Let w_j be the basis element of M_j, such that

$$\rho_j(\tilde{\gamma})w_j = (1+T)w_j. \tag{41}$$

Define

$$V = \bigoplus_{j=1}^{n} V_j, \quad M = \bigoplus_{j=1}^{n} M_j, \tag{42}$$

and

$$\rho = \bigoplus_{j=1}^{n} \rho_j : \mathrm{Gal}(\bar{\mathbb{Q}}/\mathbb{Q}) \to \bigoplus_{j=1}^{n} \mathrm{Aut}_{R_{\mathcal{P}}}(M_j), \tag{43}$$

and consider the element w in M defined as

$$w = w_1 + w_2 \cdots + w_n. \tag{44}$$

We now consider the image, which we denote by $\Phi_{\mathcal{P}}$, of the localized Hecke algebra $\mathbb{T}_{i,\mathcal{P}}$ in $\prod_{j=1}^{n} R_{\mathcal{P}}$ under the injective map (35). Note that the image of the Hecke operator T_p is a unit under (35) because the f_j are ordinary eigenforms. Hence, by (37) and (31), we conclude that $\Phi_{\mathcal{P}}$ is the $R_{\mathcal{P}}$-subalgebra generated by the identity and all the

$$\left\{ \bigoplus_{j=1}^{n} \mathrm{Trace}(\rho_j(\mathrm{Frob}_l),\ l \neq p \right\}. \tag{45}$$

Similarly, we define

$$\Phi \subseteq \bigoplus_{j=1}^{n} R \tag{46}$$

to be the R-subalgebra generated by the identity and the elements (45).

Definition 8.1. $\mathfrak{R}_{\mathcal{P}} = \Phi_{\mathcal{P}}[\mathrm{Gal}(\bar{\mathbb{Q}})/\mathbb{Q})]$, $\mathfrak{R} = \Phi[\mathrm{Gal}(\bar{\mathbb{Q}}/\mathbb{Q})]$.

Of course, $\mathfrak{R}_{\mathcal{P}}$ is a $\Phi_{\mathcal{P}}$-algebra, and V and M (cf. (42)) have natural structures as $\mathfrak{R}_{\mathcal{P}}$-modules with $\mathrm{Gal}(\bar{\mathbb{Q}}/\mathbb{Q})$ acting via ρ. Clearly, the homomorphism ρ of (43) can be extended by $\Phi_{\mathcal{P}}$-linearity to a map

$$\rho : \mathfrak{R}_{\mathcal{P}} \to \bigoplus_{j=1}^{n} \mathrm{End}_{R_{\mathcal{P}}}(M_j). \tag{47}$$

Obviously $\mathfrak{R}_{\mathcal{P}}$ is non-commutative.

Definition 8.2. $\mathcal{L} = \mathfrak{R}w$ and $\mathcal{L}_{\mathcal{P}} = \mathfrak{R}_{\mathcal{P}}w$.

Recalling that $R_{\mathcal{P}}$ is a discrete valuation ring, it is clear that $\mathcal{L}_{\mathcal{P}}$ is a free $R_{\mathcal{P}}$-module of finite rank, since it is a submodule of the free $R_{\mathcal{P}}$-module M.

Lemma 8.3. *The element T does not lie in the ideal $\mathcal{P}$ of R.*

Proof. Assume T lies in $\mathcal{P}$. As $\mathcal{P}$ is a prime ideal, its intersection with $\Lambda_{\mathcal{O}}$ would then be $T\Lambda_{\mathcal{O}}$. Thus as we have assumed that $\mathrm{ord}_{\mathcal{P}}(A_0^{(i)}(T)) > 0$, it would follow that $A_0^{(i)}(0) = 0$, and hence $L(0, \omega^i) = 0$. But by (10),

$$A_0^{(i)}(0) = L(\mathbb{Q}, 0, \omega^i),$$

and the complex L-value on the right hand side is well-known to be non-zero. This contradiction completes the proof of the lemma. □

Definition 8.4. *The elements ε_1 and ε_2 in $\mathfrak{R}_{\mathcal{P}}$ are defined as*

$$\varepsilon_1 = \frac{-1}{T}(\tilde{\gamma} - (1+T)), \quad \varepsilon_2 = \frac{1}{T}(\tilde{\gamma} - 1).$$

Obviously, we have

$$\varepsilon_1 + \varepsilon_2 = 1. \tag{48}$$

Also,

$$\rho(\varepsilon_1) = \bigoplus_j \begin{pmatrix} 1 & 0 \\ 0 & 0 \end{pmatrix}, \quad \rho(\varepsilon_2) = \bigoplus_j \begin{pmatrix} 0 & 0 \\ 0 & 1 \end{pmatrix}. \tag{49}$$

The following lemma is then obvious:

Lemma 8.5. *For all v in $\mathcal{L}_{\mathcal{P}}$, we have*

$$\varepsilon_i^2 v = \varepsilon_i v \ \ (i = 1, 2), \ \ \varepsilon_1\varepsilon_2 v = \varepsilon_2\varepsilon_1 v = 0. \tag{50}$$

□

Definition 8.6. $\mathcal{L}_{1,\mathcal{P}} = \varepsilon_1 \mathcal{L}_{\mathcal{P}}$, $\mathcal{L}_{2,\mathcal{P}} = \varepsilon_2 \mathcal{L}_{\mathcal{P}}$.

Plainly $\mathcal{L}_{1,\mathcal{P}}$ and $\mathcal{L}_{2,\mathcal{P}}$ are $\Phi_{\mathcal{P}}$-modules such that $\mathcal{L}_{\mathcal{P}} = \mathcal{L}_{1,\mathcal{P}} \oplus \mathcal{L}_{2,\mathcal{P}}$. As $\Phi_{\mathcal{P}}$ is commutative, the $\Phi_{\mathcal{P}}$-Fitting ideals of both $\mathcal{L}_{1,\mathcal{P}}$ and $\mathcal{L}_{2,\mathcal{P}}$ are defined (see [12]).

Lemma 8.7. *The $\Phi_{\mathcal{P}}$-module $\mathcal{L}_{2,\mathcal{P}}$ is free of rank one, generated by w. In particular, the $\Phi_{\mathcal{P}}$-Fitting ideal of $\mathcal{L}_{2,\mathcal{P}}$ is zero.*

Proof. First note that $\Phi_{\mathcal{P}} w = \Phi_{\mathcal{P}} \varepsilon_2 w$, since $\varepsilon_2 w = w$. Hence $\Phi_{\mathcal{P}} w \subseteq \mathcal{L}_{2,\mathcal{P}} = \varepsilon_2 \mathcal{L}_{\mathcal{P}}$. To prove the converse, for r in $\mathfrak{R}$, we write

$$\rho(r) = \bigoplus_j \begin{pmatrix} a_{r,j} & b_{r,j} \\ c_{r,j} & d_{r,j} \end{pmatrix} \in \bigoplus_j M_2(R_{\mathcal{P}}),$$

whence

$$\rho(\varepsilon_2 r) w = \sum_{j=1}^{n} d_{r,j} w_j.$$

But $d_{r,j} = \operatorname{Trace} \rho_j(\varepsilon_2 r)$, so that $\oplus_j d_{r,j}$ is in $\Phi_{\mathcal{P}}$. Thus $\varepsilon_2 \mathfrak{R}_{\mathcal{P}} w$ is contained in $\Phi_{\mathcal{P}} w$, proving that $\mathcal{L}_{2,\mathcal{P}} = \Phi_{\mathcal{P}} w$.

To show that $\mathcal{L}_{2,\mathcal{P}}$ is free as a $\Phi_{\mathcal{P}}$-module, suppose $t \neq 0$ is an element of $\Phi_{\mathcal{P}}$ such that $t\mathcal{L}_{2,\mathcal{P}} = 0$. Then $\rho_j(t\varepsilon_2 r) w_j = 0$ for all r in $\mathfrak{R}_{\mathcal{P}}$ and $1 \leq j \leq n$. Let j be such that the j-th component of t is non-zero. Then we must have $\rho_j(t\varepsilon_2 r) w_j = 0$ for all r in $\mathfrak{R}_{\mathcal{P}}$. But then $\rho_j(\varepsilon_2 r)$ annihilates the whole of M_j. By (49), this clearly implies that ρ_j is reducible as a representation of $\mathrm{Gal}(\bar{\mathbb{Q}}/\mathbb{Q})$, which contradicts the irreducibility of the ρ_j's. Finally, as the $\Phi_{\mathcal{P}}$-annihilator if $\mathcal{L}_{2,\mathcal{P}}$ is zero, its $\Phi_{\mathcal{P}}$-Fitting ideal is also zero (see [12]). This completes the proof. □

Recall that we have extended ρ to a $\Phi_{\mathcal{P}}$-linear homomorphism from $\mathfrak{R}_{\mathcal{P}}$ to $\oplus \operatorname{End}_{R_{\mathcal{P}}}(M_j)$ (47). For each r in $\mathfrak{R}_{\mathcal{P}}$, we can restrict $\rho(r)$ to obtain a $\Phi_{\mathcal{P}}$-linear endomorphism of $\mathcal{L}_{\mathcal{P}} = \mathfrak{R}_{\mathcal{P}} w$, which we denote by $\tilde{\rho}(r)$. Since $\mathcal{L}_{1,\mathcal{P}}$ and $\mathcal{L}_{2,\mathcal{P}}$ are $\Phi_{\mathcal{P}}$-modules with $\mathcal{L}_{\mathcal{P}} = \mathcal{L}_{1,\mathcal{P}} \oplus \mathcal{L}_{2,\mathcal{P}}$, we can therefore write

$$\tilde{\rho}(r) = \begin{pmatrix} \mathcal{A}_r & \mathcal{B}_r \\ \mathcal{C}_r & \mathcal{D}_r \end{pmatrix}, \tag{51}$$

where $\mathcal{A}_r$ belongs to $\operatorname{Hom}_{\Phi_{\mathcal{P}}}(\mathcal{L}_{1,\mathcal{P}}, \mathcal{L}_{1,\mathcal{P}})$, $\mathcal{B}_r \in \operatorname{Hom}_{\Phi_{\mathcal{P}}}(\mathcal{L}_{2,\mathcal{P}}, \mathcal{L}_{1,\mathcal{P}})$, $\mathcal{C}_r \in \operatorname{Hom}_{\Phi_{\mathcal{P}}}(\mathcal{L}_{1,\mathcal{P}}, \mathcal{L}_{2,\mathcal{P}})$, $\mathcal{D}_r \in \operatorname{Hom}_{\Phi_{\mathcal{P}}}(\mathcal{L}_{2,\mathcal{P}}, \mathcal{L}_{2,\mathcal{P}})$. Note that by the previous lemma, $\operatorname{Hom}_{\Phi_{\mathcal{P}}}(\mathcal{L}_{2,\mathcal{P}}, \mathcal{L}_{2,\mathcal{P}}) \simeq \Phi_{\mathcal{P}}$.

We define $\mathcal{I}_{i,\mathcal{P}}$ to be the ideal of $\Phi_{\mathcal{P}}$ which is the image of $\mathbb{I}_i \mathbb{T}_{i,\mathcal{P}}$ under the injection (35). Note that $\mathcal{I}_{i,\mathcal{P}}$ is the ideal of $\Phi_{\mathcal{P}}$ generated by all elements of the form

$$\bigoplus_j (\operatorname{Trace} \rho_j(\mathrm{Frob}_l) - 1 - \theta_i(\mathrm{Frob}_l)). \tag{52}$$

Here we recall that

$$\theta_i(\mathrm{Frob}_l) = \omega^i(l)(1+T)^{e_l}.$$

Lemma 8.8. *The following assertions hold:*

(i) $\mathcal{B}_\sigma = 0$ *for all* σ *in the decomposition group* D_p.

(ii) *The* $\Phi_{\mathcal{P}}$*-submodule of* M *generated by the images of all the maps* $\mathcal{B}_r$*, with* r *ranging over* $\mathfrak{R}$ *is equal to* $\mathcal{L}_{1,\mathcal{P}}$.

(iii) *For* σ *in* $\mathrm{Gal}(\bar{\mathbb{Q}}/\mathbb{Q})$, $\mathcal{D}_\sigma \equiv \theta_i(\sigma) \bmod \mathcal{I}_{i,\mathcal{P}}$*, where* θ_i *is given by* (38)*, and we view* $\mathcal{D}_\sigma$ *as an element of* $\Phi_{\mathcal{P}}$.

(iv) *For each* r *in* $\mathfrak{R}$*, the image of* $\mathcal{C}_r$ *is contained in* $\mathcal{I}_{i,\mathcal{P}}w$.

(v) *For* σ *in* $\mathrm{Gal}(\bar{\mathbb{Q}}/\mathbb{Q})$*, we have*

$$\tilde{\rho}(\sigma) \equiv \begin{pmatrix} 1 & \mathcal{B}_\sigma \\ 0 & \theta_i(\sigma) \end{pmatrix} \bmod \mathcal{I}_{i,\mathcal{P}}. \tag{53}$$

(vi) $\mathcal{L}_{1,\mathcal{P}}/\mathcal{I}_{i,\mathcal{P}}\mathcal{L}_{1,\mathcal{P}}$ *is stable under* $\mathrm{Gal}(\bar{\mathbb{Q}}/\mathbb{Q})$ *and this action is trivial.*

Proof. Assertion (i) is clear from (39). Turning to (ii), we note that for each r in $\mathfrak{R}$, we have

$$rw \in \mathrm{Im}\,\mathcal{B}_r \oplus \mathrm{Im}\,\mathcal{D}_r \subseteq \mathcal{L}_{1,\mathcal{P}} \oplus \mathcal{L}_{2,\mathcal{P}}.$$

Indeed, this is clear if r belongs to $\Phi_{\mathcal{P}}$, and it is also plain for r in $\mathrm{Gal}(\bar{\mathbb{Q}}/\mathbb{Q})$, since the component w_j of w is an element of our fixed $R_{\mathcal{P}}$-basis of M_j. Also, since $\mathcal{D}_r$ is isomorphic to $\Phi_{\mathcal{P}}$ by the previous lemma, we conclude that the image of $\mathcal{D}_r = \mathcal{L}_{2,\mathcal{P}}$ for any r in $\mathfrak{R}_{\mathcal{P}}$. Assertion (ii) now follows from the fact that $\mathcal{L}_{\mathcal{P}} = \mathfrak{R}_{\mathcal{P}}.w$.
We next prove (iii). For each r in $\mathfrak{R}_{\mathcal{P}}$, we have by (52),

$$\bigoplus_j (\mathrm{Trace}\,\rho_j(\varepsilon_2 r)) \equiv (\mathbf{1} \oplus \theta_1)(\varepsilon_2 r) \bmod \mathcal{I}_{i,\mathcal{P}}, \tag{54}$$

where $\mathbf{1}$ denotes the trivial character. The left hand side of (54) is easily seen to be $\mathcal{D}_r$, which we continue to view as an element of $\Phi_{\mathcal{P}}$. On the other hand, we have

$$\mathbf{1}(\varepsilon_2) = 0, \quad \theta_i(\varepsilon_2) = 1.$$

Hence the right hand side of (54) is equal to $\theta_i(r)$, establishing (iii).
Turning to (iv), note that we have the identity

$$\mathcal{D}_{rr'} = \mathcal{D}_r\mathcal{D}_{r'} + \mathcal{C}_r\mathcal{B}_{r'}$$

for all elements r, r' in $\mathfrak{R}_{\mathcal{P}}$. Since θ_i is a character, it follows from (iii) that

$$\mathrm{Im}\,\mathcal{C}_r\mathcal{B}_{r'} \subseteq \mathcal{I}_{i,\mathcal{P}}\mathcal{L}_{2,\mathcal{P}}.$$

As the image of the $\mathcal{B}_{r'}$ fills up the whole of $\mathcal{L}_{1,\mathcal{P}}$ as r' varies, we conclude that $\mathrm{Im}\,\mathcal{C}_r \subset \mathcal{I}_{i,\mathcal{P}}\mathcal{L}_{2,\mathcal{P}}$, proving (iv).
To establish (v) and (vi), we must show that $A_\sigma \bmod \mathcal{I}_{i,\mathcal{P}}$ is the identity map for any σ in $\mathrm{Gal}(\bar{\mathbb{Q}}/\mathbb{Q})$. Suppose

$$\rho(\sigma) = \oplus \begin{pmatrix} a_{\sigma,j} & b_{\sigma,j} \\ c_{\sigma,j} & d_{\sigma,j} \end{pmatrix} \text{ in } \bigoplus_{j=1}^{n} \mathrm{GL}_2(R_{\mathcal{P}}).$$

The action of $\mathcal{A}_\sigma$ on the element $\varepsilon_1 rw$ in $\mathcal{L}_{1,\mathcal{P}}$ is given by multiplication by $a_\sigma = \oplus a_{\sigma,j}$. But $a_{\sigma,j} = \text{Trace}(\rho_j(\varepsilon_1\sigma))$ as

$$\rho_j(\varepsilon_1\sigma) = \begin{pmatrix} a_{\sigma,j} & b_{\sigma,j} \\ 0 & 0 \end{pmatrix}.$$

On the other hand, by (52), we have

$$a_{\sigma,j} \equiv \text{Trace}(\mathbf{1} + \theta_i)(\varepsilon_1\sigma) = 1 \bmod \mathcal{I}_{i,\mathcal{P}},$$

because

$$\theta_i(\varepsilon_1) = 0, \quad \mathbf{1}(\varepsilon_1) = 1.$$

This completes the proof. □

9. Completion of the proof

We continue to assume that i is an odd integer with $i \not\equiv -1 \bmod p-1$. We now give an equivalent form of (29) which fits in better with our present arguments. Define R_i to be R endowed with the following action of $\text{Gal}(\bar{\mathbb{Q}}/\mathbb{Q})$:

$$\sigma(r) = \theta_i(\sigma)r,$$

where θ_i is the character (38). The Pontryagin dual

$$\overset{\vee}{R}_i = \text{Hom}\,(R_i, \mathbb{Q}_p/\mathbb{Z}_p)$$

is therefore a discrete p-primary abelian group on which $\text{Gal}(\bar{\mathbb{Q}}/\mathbb{Q})$ acts via θ_i^{-1}. For each subfield $\mathcal{K}$ of $\bar{\mathbb{Q}}$, define the subgroup of unramified cocycles by

$$H^1_{\text{nr}}(\mathcal{K}, \overset{\vee}{R}_i) = \text{Ker}\left(H^1(\text{Gal}(\bar{\mathbb{Q}}/\mathcal{K}), \overset{\vee}{R}_i) \to \prod_v H^1(I_v, \overset{\vee}{R}_i)\right),$$

where the product is taken over all non-archimedean primes v of $\mathcal{K}$, and I_v denotes the inertial subgroup of some fixed prime of $\bar{\mathbb{Q}}$ above v. We recall that $X_\infty = \text{Gal}(\mathcal{L}_\infty/\mathcal{F}_\infty)$, where $\mathcal{L}_\infty$ is the maximal unramifed abelian p-extension of $\mathcal{F}_\infty$. As is explained at the beginning of §3, there is a natural action of $\mathcal{G} = \Delta \times \Gamma$ on X_∞. We define

$$V_\infty = X_\infty^\bullet,$$

where as in §3, the dot means that the $\mathcal{G}$-action on X_∞ has been inverted. Define

$$\text{Sel}_i(R) = H^1_{\text{nr}}(\text{Gal}(\bar{\mathbb{Q}}/\mathbb{Q}), \overset{\vee}{R}_i), \quad Z_i(R) = \text{Hom}\,(\text{Sel}_i(R), \mathbb{Q}_p/\mathbb{Z}_p). \tag{55}$$

Lemma 9.1.

$$Z_i(R) = V_\infty^{(i)} \otimes_{\Lambda(\Gamma)} R.$$

Proof. Note that $(\overset{\vee}{R}_i)^\Delta = 0$ since i is not congruent to zero modulo $p-1$. Hence

$$H^j(\mathcal{G}, \overset{\vee}{R}_i) = H^j(\Gamma, (\overset{\vee}{R}_i)^\Delta) = 0 \quad (j \geq 0). \tag{56}$$

Thus

$$H^1(\mathrm{Gal}(\bar{\mathbb{Q}}/\mathbb{Q}), \check{R}_i) \simeq H^1(\mathrm{Gal}(\bar{\mathbb{Q}}/\mathcal{F}_\infty), \check{R}_i)^{\mathcal{G}}.$$

As p is totally ramified in $\mathbb{Q}(\mu_{p^\infty})$, one sees easily that this induces an isomorphism

$$\mathrm{Sel}_i(R) \simeq H^1_{\mathrm{nr}}(\mathrm{Gal}(\bar{\mathbb{Q}}/\mathcal{F}_\infty), \check{R}_i)^{\mathcal{G}} = \mathrm{Hom}_{\mathcal{G}}(X_\infty, \check{R}_i).$$

But recalling that $\mathcal{G}$ acts on $\check{R}_i$ by θ_i^{-1}, we see that

$$\mathrm{Hom}_{\mathcal{G}}(X_\infty, \check{R}_i) = \mathrm{Hom}_{\Lambda(\Gamma)}(V_\infty^{(i)}, \check{R}_i).$$

But

$$\begin{aligned} \mathrm{Hom}_{\Lambda(\Gamma)}(V_\infty^{(i)}, \check{R}_i) &= \mathrm{Hom}_{\Lambda(\Gamma)}(V_\infty^{(i)}, \\ \mathrm{Hom}\,(R, \mathbb{Q}_p/\mathbb{Z}_p)) &= \mathrm{Hom}\,(V_\infty^{(i)} \otimes_{\Lambda(\Gamma)} R, \mathbb{Q}_p/\mathbb{Z}_p), \end{aligned}$$

and this completes the proof. □

Since $V_\infty^{(i)}$ is $X_\infty^{(-i)}$ with its Γ-action inverted, we conclude that, in order to prove (29), it suffices to show that for every prime ideal $\mathcal{P}$ of R of height one, we have

$$\mathrm{ord}_{\mathcal{P}}\,(\mathrm{char}_R\, Z_i(R)) \geq \mathrm{ord}_{\mathcal{P}}\,(A_0^{(i)}(T)). \tag{57}$$

Define

$$\begin{aligned} \mathcal{N} &= \mathrm{Im}(\mathcal{L} \to \mathcal{L}_{\mathcal{P}}/\mathcal{I}_{i,\mathcal{P}}\mathcal{L}_{\mathcal{P}}), \\ \mathcal{N}_1 &= \mathcal{N} \cap \mathcal{L}_{1,\mathcal{P}}/\mathcal{I}_{i,\mathcal{P}}\mathcal{L}_{1,\mathcal{P}} \quad \text{and} \quad \mathcal{N}_2 = \mathcal{N} \cap \mathcal{L}_{2,\mathcal{P}}/\mathcal{I}_{i,\mathcal{P}}\mathcal{L}_{2,\mathcal{P}}. \end{aligned} \tag{58}$$

If V is any $\mathrm{Gal}(\bar{\mathbb{Q}}/\mathbb{Q})$-module, which is also an R-module, we write $V(\theta_i^{-1})$ for V with the new Galois action defined by

$$\sigma \circ v = \theta_i^1(\sigma) \cdot \sigma v,$$

where the action σv on the right is the old Galois action. Put

$$\mathfrak{X} = \mathrm{Hom}_R(\mathcal{N}_1(\theta_i^{-1}), \check{R}_i).$$

Since $\mathrm{Gal}(\bar{\mathbb{Q}}/\mathbb{Q})$ acts trivially on $\mathcal{N}_1$, it is clear that each β in $\mathfrak{X}$ is a $\mathrm{Gal}(\bar{\mathbb{Q}}/\mathbb{Q})$-homomorphism. Given any β in $\mathfrak{X}$, we define a map

$$c_\beta : \mathrm{Gal}(\bar{\mathbb{Q}}/\mathbb{Q}) \to \check{R}_i$$

by the formula

$$c_\beta(\sigma) = \beta(\mathcal{B}_\sigma(w) \cdot \theta_i(\sigma)^{-1} \bmod \mathcal{I}_{i,\mathcal{P}}), \tag{59}$$

where $\mathcal{B}_\sigma$ is the homomorphism given in (51), and hence the right hand side of (59) does indeed lie in $\beta(\mathcal{N}_1)$. For σ, τ in $\mathrm{Gal}(\bar{\mathbb{Q}}/\mathbb{Q})$, we conclude from (53) that

$$\mathcal{B}_{\sigma\tau}(w) = (\mathcal{B}_\tau(w) + \mathcal{B}_\sigma(w)\,\theta_i(\tau)) \bmod \mathcal{I}_{i,\mathcal{P}}.$$

Recalling that $\mathrm{Gal}(\bar{\mathbb{Q}}/\mathbb{Q})$ acts on $\overset{\vee}{R}_i$ via θ_i^{-1}, it follows immediately that c_β is in fact a 1-cocycle, and we denote its class in $H^1(\mathrm{Gal}(\bar{\mathbb{Q}}/\mathbb{Q}), \overset{\vee}{R}_i)$ by $\mathfrak{c}_\beta$. We can therefore define a homomorphism

$$\Omega : \mathfrak{X} \to H^1(\mathrm{Gal}(\bar{\mathbb{Q}}/\mathbb{Q}), \overset{\vee}{R}_i) \tag{60}$$

by $\Omega(\beta) = \mathfrak{c}_\beta$.

Lemma 9.2. *The image of* Ω *is contained in* $H^1_{\mathrm{nr}}(\mathrm{Gal}(\bar{\mathbb{Q}}/\mathbb{Q}, \overset{\vee}{R}_i) = \mathrm{Sel}_i(R)$.

Proof. Since the Galois representations ρ_j (see (36)) are unramified outside p, and $\mathcal{B}_\sigma = 0$ for all σ in the decomposition group D_p by Lemma (8.8), assertion (i) is clear. □

Proposition 9.3. *The map* Ω *is injective after tensoring with the localisation* $R_{\mathcal{P}}$ *of* R.

Proof. If β is any element of $\mathfrak{X}$, we have the extension of $\mathrm{Gal}(\bar{\mathbb{Q}}/\mathbb{Q})$-modules

$$0 \to \mathcal{N}_1(\theta_i^{-1}) \to \mathcal{N}(\theta_i^{-1}) \to \mathcal{N}_2(\theta_i^{-1}) \to 0,$$

which clearly further gives rise to the extension

$$0 \to \mathcal{N}_1(\theta_i^{-1})/\mathrm{Ker}\ \beta \to \mathcal{N}(\theta_i^{-1})/\mathrm{Ker}\ \beta \to \mathcal{N}_2(\theta_i^{-1}) \to 0. \tag{61}$$

Since $\mathcal{N}_2$ can be identified with the module $(R/J_i) \cdot w$, the cohomology class of the extension (61) in $H^1(\mathrm{Gal}(\bar{\mathbb{Q}}/\mathbb{Q}), \mathcal{N}_1(\theta_i^{-1})/\mathrm{Ker}\ \beta)$ is given by the class of the cocycle

$$d(\sigma) = \theta_i(\sigma)^{-1}\sigma w - w \bmod (\mathrm{Ker}\ \beta). \tag{62}$$

Obviously, we have

$$\beta(d(\sigma)) = c_\beta(\sigma). \tag{63}$$

Assume now that β in $\mathfrak{X}$ is such that c_β is a coboundary, i.e. there exists t in $\overset{\vee}{R}_i$ such that

$$c_\beta(\sigma) = (\theta_i(\sigma)^{-1} - 1)t \quad \text{for all } \sigma \text{ in } \mathrm{Gal}(\bar{\mathbb{Q}}/\mathbb{Q}).$$

Since i is odd, we can choose σ_0 in $\mathrm{Gal}(\bar{\mathbb{Q}}/\mathbb{Q})$ such that $(\theta_i(\sigma_0)^{-1} - 1)$ is a unit in R. Hence

$$\beta(s) = t, \quad \text{where} \quad s = (\theta_i(\sigma_0)^{-1} - 1)^{-1} d(\sigma_0).$$

It follows easily that d itself is a coboundary. Hence the extension (61) is split, and so there exists a $\mathrm{Gal}(\bar{\mathbb{Q}}/\mathbb{Q})$-surjection

$$\mathcal{N}(\theta_i^{-1})/\mathrm{Ker}\ \beta \to \mathcal{N}_1(\theta_i^{-1})/\mathrm{Ker}\ \beta. \tag{64}$$

Recalling that $\mathcal{N}$ is cyclic and generated by w, it follows that the module on the right of (64) is cyclic with generator x, say. By virtue of (41), it follows that the module on the left of (64) is annihilated by $(\tilde{\gamma} - 1)$, and so x is also annihilated by $(\tilde{\gamma} - 1)$.

As $\mathrm{Gal}(\bar{\mathbb{Q}}/\mathbb{Q})$ acts trivially on $\mathcal{N}_1$, we conclude that x is annihilated by $((1+T)^{-1}-1)$ and so also $T \cdot x = 0$. Hence

$$T\mathcal{N}_1(\theta_i^{-1}) \subset \text{Ker } \beta. \tag{65}$$

We now define

$$\mathfrak{X}' = \{\beta \text{ in } \mathfrak{X} \text{ such that } c_\beta \text{ is a coboundary}\}.$$

Under the natural pairing

$$\mathcal{N}_1 \times \mathfrak{X} \to \overset{\vee}{R_i},$$

$\mathfrak{X}'$ is dual to $\mathcal{N}_1(\theta_i^{-1})/W$, where

$$W = \bigcap_{\beta \in \mathfrak{X}'} \text{Ker } \beta.$$

However by (65),

$$W \supset T\mathcal{N}_1(\theta_i^{-1}).$$

Thus $T \cdot \mathfrak{X}' = 0$, whence $\mathfrak{X}'_{\mathcal{P}} = 0$, since $T \notin \mathcal{P}$. This proves that the map Ω is injective after tensoring with $R_{\mathcal{P}}$ over R and the proof of the proposition is complete. □

Note that as an R-module, we have

$$\mathfrak{X} = \text{Hom}_R(\mathcal{N}_1(\theta_i^{-1}), \overset{\vee}{R_i}) = \text{Hom}(\mathcal{N}_1, \mathbb{Q}_p/\mathbb{Z}_p),$$

and so, dualizing the map Ω, we obtain a homomorphism of R-modules

$$\overset{\vee}{\Omega} : Z_i(R) \twoheadrightarrow \mathcal{N}_1,$$

which is surjective after localisation at $\mathcal{P}$ by Proposition 9.3. Thus, for every prime ideal $\mathcal{P}$ of height one of R, we have

$$\text{ord}_{\mathcal{P}}\, \text{char}_{R_{\mathcal{P}}}(Z_i(R)_{\mathcal{P}}) \geq \text{ord}_{\mathcal{P}}\, \text{char}_{R_{\mathcal{P}}}((\mathcal{N}_1)_{\mathcal{P}}). \tag{66}$$

We recall that, by (32) and (35), we have the isomorphism of $R_{\mathcal{P}}$-algebras

$$R_{\mathcal{P}}/J_{i,\mathcal{P}} \simeq \Phi_{\mathcal{P}}/\mathcal{I}_{i,\mathcal{P}}\Phi_{\mathcal{P}}. \tag{67}$$

Using this isomorphism, we can view the $\Phi_{\mathcal{P}}$-module $\mathcal{L}_{1,\mathcal{P}}/\mathcal{I}_{\mathcal{P}}\mathcal{L}_{1,\mathcal{P}}$ as a module over $R_{\mathcal{P}}$ which is annihilated by $J_{i,\mathcal{P}}$. By arguments from commutative algebra, we shall then prove at the end of this section the following key lemma.

Lemma 9.4. *For every prime ideal $\mathcal{P}$ of height 1 of R, the $R_{\mathcal{P}}$-Fitting ideal of $\mathcal{L}_{1,\mathcal{P}}/\mathcal{I}_{i,\mathcal{P}}\,\mathcal{L}_{1,\mathcal{P}}$ is contained in $J_{i,\mathcal{P}}$.*

Assuming this lemma, we shall complete the proof of (57), and hence also the proof of the main conjecture. As the $R_{\mathcal{P}}$-Fitting ideal of $\mathcal{L}_{1,\mathcal{P}}/\mathcal{I}_{i,\mathcal{P}}\,\mathcal{L}_{1,\mathcal{P}}$ is contained in $J_{i,\mathcal{P}}$ Proposition 17 of [12] allows us to conclude that

$$\text{char}_{R_{\mathcal{P}}}(\mathcal{L}_{1,\mathcal{P}}/\mathcal{I}_{i,\mathcal{P}}\mathcal{L}_{1,\mathcal{P}}) \subseteq J_{i,\mathcal{P}}. \tag{68}$$

Combining this last inclusion with (66), and noting that

$$(\mathcal{N}_1)_{\mathcal{P}} = \mathcal{L}_{1,\mathcal{P}}/\mathcal{I}_{i,\mathcal{P}}, \mathcal{L}_{1,\mathcal{P}}$$

we have

$$\mathrm{ord}_{\mathcal{P}}\,(Z_i(R)_{\mathcal{P}}) \geq \mathrm{ord}_{\mathcal{P}}\, J_{i,\mathcal{P}}. \tag{69}$$

Recall that the Hecke algebra $\mathbb{T}_{i,\mathcal{P}}$ is isomorphic to $\Phi_{\mathcal{P}}$ and hence by (67) and Theorem 4.4, we have

$$\mathrm{ord}_{\mathcal{P}}\, J_{i,\mathcal{P}} := \mathrm{ord}_{\mathcal{P}}(\mathrm{char}(R_{\mathcal{P}}/J_{i,\mathcal{P}}) \geq \mathrm{ord}_{\mathcal{P}}\, A_0^{(i)}(T), \tag{70}$$

which proves (57) and therefore the main conjecture.

We are very grateful to F. Nuccio for providing us with the following proof of Lemma 9.4. Recall that there is a natural ring homomorphism from $R_{\mathcal{P}}$ to $\Phi_{\mathcal{P}}$ making $\Phi_{\mathcal{P}}$ an $R_{\mathcal{P}}$ algebra and note that $\Phi_{\mathcal{P}}$ is a flat $R_{\mathcal{P}}$-module as it is free of finite rank over $R_{\mathcal{P}}$. On the other hand, defining $L_1 \subset \mathcal{L}_1$ to be the R-submodule $L_1 = \varepsilon_1\, R \cdot w$, so that

$$L_{1,\mathcal{P}} \otimes_{\Phi,\mathcal{P}} \Phi_{\mathcal{P}} = \mathcal{L}_{1,\mathcal{P}}.$$

Since $\Phi_{\mathcal{P}}$ is a flat $R_{\mathcal{P}}$-module, and $L_{1,\mathcal{P}}$ is $R_{\mathcal{P}}$-faithful, it follows from [2, Chap. 1, §2, Cor. 2], that

$$\mathrm{ann}_{\Phi,\mathcal{P}}(\mathcal{L}_{1,\mathcal{P}}) = 0.$$

Hence by Corollary 14 of [12], we have that the $\Phi_{\mathcal{P}}$ -Fitting ideal satisfies

$$\mathrm{Fitt}_{\Phi_{\mathcal{P}}}(\mathcal{L}_{1,\mathcal{P}}/\mathcal{I}_{i,\mathcal{P}}\mathcal{L}_{1,\mathcal{P}}) \subset \mathcal{I}_{i,\mathcal{P}}.$$

Hence, by [12, Lemma 10], we obtain

$$\mathrm{Fitt}_{\Phi_{\mathcal{P}}/\mathcal{I}_{i,\mathcal{P}}}(\mathcal{L}_{1,\mathcal{P}}/\mathcal{I}_{i,\mathcal{P}}\mathcal{L}_{1,\mathcal{P}}) = 0.$$

Invoking the ring isomorphism (67), we conclude that

$$\mathrm{Fitt}_{R_{\mathcal{P}}/J_{i,\mathcal{P}}}(\mathcal{L}_{1,\mathcal{P}}/\mathcal{I}_{i,\mathcal{P}}\mathcal{L}_{1,\mathcal{P}}) = 0.$$

Since $\mathcal{L}_{1,\mathcal{P}}$ is a faithful $R_{\mathcal{P}}$-module, we can apply Lemma 10 of [12] once more to complete the proof of Lemma 9.4. □

References

[1] D. Banerjee, E. Ghate and V. G. Narasimha Kumar, Λ-adic forms and the Iwasawa Main Conjecture (this volume).

[2] N. Bourbaki, Elements of Mathematics, Commutative Algebra, Chapters 1–7, Springer (1989).

[3] P. Cassou-Nogues, Valeurs aux entiers négatifs des fonctions zêta p-adiques, *Invent. Math.*, **51** (1979) 29–59.

[4] J. Coates and R. Sujatha, Cyclotomic fields and zeta values, Springer Monographs in Mathematics, Springer (2006).

[5] P. Deligne and K. Ribet, Values of abelian L-functions at negative integers over totally real fields, *Invent. Math.*, **59** (1980) 227–286.

[6] H. Hida, Modules of congruence of Hecke algebras and L-functions associated with cusp forms, *Amer. J. Math.*, **110** (1988) 323–382.

[7] H. Hida, Elementary theory of L-functions and Eisenstein series, London Mathematical Society Student Texts **26**, Cambridge University Press, Cambridge (1993).

[8] K. Iwasawa, On $\mathbb{Z}_l$-extensions of algebraic number fields, *Ann. Math.*, **98** (1973) 246–326.

[9] K. Iwasawa, Lectures on p-adic L-functions, *Ann. Math. Studies*, **74**, Princeton University Press (1972).

[10] B. Mazur and A. Wiles, Class fields of abelian extensions of $\mathbb{Q}$, *Invent. Math.*, **76** (1984) 179–330.

[11] B. Mazur and A. Wiles, On p-adic analytic families of Galois representations, *Comp. Math.*, **59** (1986) 231–264.

[12] F. Nuccio, Fitting Ideals (this volume).

[13] K. Ribet, A modular construction of unramified p-extensions of $\mathbb{Q}(\mu_p)$, *Invent. Math.*, **34** (1976) 151–162.

[14] K. Rubin, The Main Conjecture Appendix in S. Lang Cyclotomic fields I and II, Combined second edition, Graduate Texts in Mathematics, 121, Springer-Verlag, New York (1990).

[15] C. Seigel, Berechnung von Zeta-funktionen an ganzzahligen Stellen, *Nach. Akad. Wiss. Göttingen*, (1969) 87–102.

[16] A. Wiles, On ordinary Λ-adic representations associated to modular foms, *Invent. Math.*, **94** (1988) 529–573.

[17] A. Wiles, The Iwasawa conjecture for totally real fields, *Ann. Math.*, **131** (1990) 493–540.

Made in the USA
Monee, IL
13 September 2026

62979107R00083